AN INTRODUCTION TO HISTORICAL PLANT GEOGRAPHY

AN INTRODUCTION TO HISTORICAL PLANT GEOGRAPHY

BY

E. V. WULFF

Curator of the Herbarium, Department of Geography of Cultivated Plants, U.S.S.R. Institute of Plant Industry, Leningrad

authorised translation by Elizabeth Brissenden

Foreword by ELMER D. MERRILL

Administrator of Botanical Collections, Harvard University

GYAN

Published by

Gyan Publishing House
5, Ansari Road
Daryaganj, New Delhi-110002
Phone: 011-47034999, 9811692060
E-mail: books@gyanbooks.com

Distribution Network
gyanbooks.com
India, USA, Canada, UK, Australia, France

ISBN : 978-81-212-2235-8 (HB)
First Published, 1943

2nd Impression 2021

Printed at: Gyan Press, Delhi.

An Introduction to Historical Plant Geography
Author: E. V. Wulff

AN INTRODUCTION TO
HISTORICAL PLANT GEOGRAPHY

BY

E. V. WULFF

Curator of the Herbarium, Department of Geography of Cultivated Plants, U.S.S.R. Institute of Plant Industry, Leningrad

authorized translation by ELIZABETH BRISSENDEN

Foreword by ELMER D. MERRILL
Administrator of Botanical Collections, Harvard University

WALTHAM, MASS., U.S.A.

Published by the Chronica Botanica Company

A NEW SERIES OF PLANT SCIENCE BOOKS

edited by Frans Verdoorn

Volume X

HISTORICAL PLANT GEOGRAPHY

UNIFORM WITH THIS VOLUME

MacDougal — Tree Growth
Grant — Wood Pulp
Darrah — Principles of Paleobotany
Pfeiffer — Experimentelle Cytologie
Bawden — Plant Viruses and Virus Diseases
Guilliermond — The Cytoplasm of the Plant Cell
Reed — A Short History of the Plant Sciences
Baldwin — Forest Tree Seed
Lloyd — The Carnivorous Plants

IN PREPARATION

Schopfer — Plants and Vitamins
Erdtman — Pollen Analysis
Hoagland — Lectures on Plant Nutrition
Browne — Source Book of Agricultural Chemistry

AN INTRODUCTION TO
HISTORICAL PLANT GEOGRAPHY

BY

E. V. WULFF

Curator of the Herbarium, Department of Geography of Cultivated Plants, U.S.S.R. Institute of Plant Industry, Leningrad

authorized translation by ELIZABETH BRISSENDEN

Foreword by ELMER D. MERRILL
Administrator of Botanical Collections, Harvard University

1943

WALTHAM, MASS., U.S.A.

Published by the Chronica Botanica Company

FOREWORD

Impressed with the significance and value of Dr. Wulff's Russian text, which was published in 1932 and again in a slightly revised edition in 1933, I then suggested to him the desirability of an English translation in order that his views might be made more generally available to a wider public. The present volume, revised and brought up to date by Dr. Wulff, as far as war conditions permitted him to do so, has been excellently translated by Miss Elisabeth Brissenden, working in close association with Dr. Wulff in Leningrad. It constitutes the first part, covering general and theoretical problems, of a projected three volume work on historical plant geography. The second part, devoted to a History of the Floras of the World, was completed and, at the outbreak of the present war, ready for publication under the auspices of the Academy of Sciences of the U. S. S. R. in Leningrad. The third part, dealing with the changes in floras caused by man's activities, is in the course of preparation. It is to be hoped that after these two volumes are published, English versions of both may appear corresponding to this one.

The present volume consists of eleven chapters opening with one covering the scope of the subject, the relationships to allied sciences, and methods of investigation, and closing with an excellently prepared one on the concept of floral elements. Between these two chapters, in much detail, is considered the history of the science, areas: their types and origins, parallelisms in the geographical distribution of plants and animals, artificial and natural factors in relation to the geographic distribution of plants, migrations of species and of floras and their causes, and the historical causes for the present structure of areas and the composition of floras.

Thus in this single volume students and investigators will find assembled in one place a great amount of well coordinated data, presented in a lucid manner, that should greatly lighten their burdens, and act as a stimulant to further investigations. Much work remains to be done and some of the theories discussed need further testing through the laborious process of assembling further details. The publication of this volume in English now makes Dr. Wulff's views very generally available to a wide public. The author, the translator, and the publisher deserve sincere thanks, for the present volume is a mine of logically and authoritatively discussed information on the subject.

Dr. Wulff's enthusiastic adherence to the Wegener hypothesis will not be accepted by all plant geographers and geologists; many, in fact, are strongly opposed to it. Although it is interesting to note that Wegener's theory has recently enjoyed support in geological circles (see Sir Thomas Holland's recent Bruce-Preller lecture on the problems of continental drift, Proc. Roy. Soc. Edinb. B, 61, II (13), p. 149 et seq. 1941), others take violent exception to it (see BAILEY WILLIS, On the Life History of Theory, Am. Scientist 30:290, 1942).

From a purely botanical standpoint I compiled certain data from the published volumes of the second edition of the ENGLER-PRANTL "Natürliche Pflanzenfamilien", in which all the known genera, approximating 2811 in all the families of the Gymnospermae, 26 families of the Monocotyledoneae, and 95 families of the Dicotyledoneae are considered. Only about 180 of these 2800 genera have species common to both hemispheres. This is a good case of random sampling as the groups considered include many families of plants that are strictly tropical, and many such as the *Pinaceae, Cupressaceae, Liliaceae, Saxifragaceae, Rosaceae, Cruciferae, Caryophyllaceae*, and others, that are very well developed in the temperate zone with a fair number of representatives even in the far north. The genera that are common to both hemispheres are but about 6.3 percent of the total. Were strictly tropical groups considered, it would be much smaller. I merely cite these figures for what they may be worth, but were the theory of drifting continents correct one would logically expect to find a great number of genera common to the tropics of both hemispheres.

The editors of CHRONICA BOTANICA are under obligations to Dr. H. M. RAUP, of the Arnold Arboretum, and to Dr. and Mrs. STANLEY A. CAIN, of the University of Tennessee, for their assistance and advice in preparing the present text for the press.

The appearance of this volume at this time emphasizes the fact that science is not limited by international boundaries and that the work of one investigator prosecuted in one country under one regime, contributes materially to the advancement of knowledge in all countries. I should like to quote a statement of Mr. RAYMOND FOSDICK in a recent report of the Rockefeller Foundation: "In the shadows that are deepening over Europe the lights of learning are fading one by one. The conception of knowledge as an international responsibility has vanished. The free flow of ideas across boundary lines between laboratories and universities had dried up. Elsewhere the exigencies of war have erased the opportunities for intellectual and cultural life as that term was understood a few years ago." And yet, in spite of the war, and in spite of the fact that the U. S. S. R. has suffered most grievously, it has been possible to provide for the English translation of the present volume in Russia, and its publication in the United States of America.

The author, Dr. E. V. WULFF, was born in 1885. He received his Ph.D. degree from the University of Vienna in 1910. He is well known for his publications dealing with historical plant geography, on the flora of Crimea, various phases of economic botany, and systematic work on the *Scrophulariaceae*. He is Curator of the Herbarium, Department of Geography of Cultivated Plants of the Institute of Plant Industry of the U. S. S. R., Leningrad.

Appended to this foreword is a statement prepared by Dr. HUGH M. RAUP of the Arnold Arboretum, in which various significant papers are listed, supplementing the numerous references given by Dr. WULFF. Most of these appeared in the interim between the publication and revision of the original Russian edition and this English translation of WULFF's work.

ELMER D. MERRILL

Administrator of Botanical Collections, Harvard University, and Director of the Arnold Arboretum

JAMAICA PLAIN, MASS., U. S. A.
CHRISTMAS, 1942

For the student of plant geography, one of the chief merits to be found in this translation is its analysis and discussion of a large amount of continental, especially Russian, literature which would not otherwise be readily available. It will be not less useful for its lucid discussions of the classic concepts upon which much of modern floristic plant geography is based. On the other hand, students will undoubtedly miss a number of references to American literature which deal with the various subjects that make up Professor Wulff's work. This is particularly true of a number of very recent papers, many of which will be found to add greatly to the argument of the book. The exigencies of transportation during war time have of course made it extremely difficult for students on two sides of the world to keep up with one another's work. It may therefore be suitable to add some notices of papers that are outstanding in the fields they represent. The following paragraphs are not to be considered in any sense complete, therefore, and are offered merely, in accordance with Dr. Wulff's desires as expressed to the editor of this series, as aids toward the completion of certain chapters of the book.

The historical and classic phases of geographic botany have had a brief but excellent treatment in "A Short History of the Plant Sciences" by H. S. Reed (Chronica Botanica Co., 1942); while an earlier American paper by E. L. Greene on "Landmarks of Botanical History" (Smithsonian Miscel. Coll., 1909), though not cited by Wulff, will yield much of value. With regard to the significance of Alexander von Humboldt as the founder of modern phytogeography, discussed by Wulff on p. 11, the reader will find further material in a recent book by Richard Hartshorne on "The Nature of Geography" (Ann. Ass. Am. Geogr. 29: 171–658, 1939; reprinted in book form with separate paging).

Some of the criteria for the determination of center of area, treated at some length by Wulff in his third chapter, were outlined many years ago in America by C. C. Adams: "Southeastern United States as a Center of Geographic Distribution of Flora and Fauna" (Biol. Bull. 3: 115–31, 1902). Adams' criteria have been critically re-examined and extended by Stanley A. Cain in a paper now in press.

The problems centering in the origin of geographic areas are, as Wulff makes clear (Chapter IV), closely allied to those of speciation. A number of papers have appeared recently in American biological literature which touch upon these problems, and since they appear to have advanced our thinking considerably, a few of the more important will be cited. Dobzhansky's book on "Genetics and the Origin of Species" (Col. U. P., 2nd ed., 1941) contains stimulating discussion of some of the geographic aspects of speciation. Results of the extensive transplant experiments carried on by the Carnegie Institution in California have far-reaching implications for American students of floristic geography. They have been published by J. Clausen, D. D. Keck and W. M. Hiesey in "Experimental Studies in the Nature of Species, I. Effect of varied environments on Western American Plants" (Carn. Inst. Wash. Publ. 520, 1940). Other aspects of the genetical approach to geographic problems are to be found in papers cited below in supplementing Wulff's treatment of migration (Chapter IX). The relation of recent studies of polyploidy to geographic problems is noted in papers by Edgar Anderson, "Cytology in its Relation to Taxonomy" (Bot. Rev. 3: 335–50, 1937), and by G. L. Stebbins, "The Significance of Polyploidy in Plant Evolution" (Am. Nat. 74: 54–56, 1940).

The chapter on "Types of Areas" (V) could be illustrated more extensively from American literature than WULFF has done. Maps prepared by M. L. FERNALD and published in various papers by him yield a wealth of material in this connection. The following are references to a few of them: "Persistence of Plants in Unglaciated Areas of Boreal America" (Mem. Gray Herb. 2, 1925); "The Antiquity and Dispersal of Vascular Plants" (Quart. Rev. Biol. 1: 212–45, 1926); "Some Relations of the Floras of the Northern Hemisphere" (Proc. Int. Cong. Pl. Sci. 2: 1487–1507, 1929); "Specific Segregations and Identities in Some Floras of Eastern North America and the Old World" (Rhodora 33: 25–63, 1931); "Recent Discoveries in the Newfoundland Flora" (Contr. Gray Herb. 101, 1933); "A Century of Additions to the Flora of Virginia" (Contr. Gray Herb. 133, 1940). The perplexing problem of species with ranges divided between North and South America has been discussed recently by G. E. DU RIETZ in "Problems of Bipolar Distribution" (Acta Phytogeog. Suecica 13: 215–82, 1940). Students will find useful data on the marginal phenomena of plant ranges, treated briefly by WULFF on p. 67, in papers by R. F. GRIGGS on his investigations of timberlines: "The Edge of the Forest in Alaska" (Ecology 15: 80–96, 1934); "Timberlines in the Northern Rocky Mountains" (Ecology 19: 548–64, 1938); "Indications as to Climatic Changes from the Timberline of Mount Washington" (Science 95: 515–19, 1942).

Around the general subject of plant migrations and their causes there has grown an imposing literature, both European and American. Students in the Americas will welcome WULFF's resumé of the European material, but will no doubt wish that circumstances had permitted him to correlate it more fully with recent American work. The whole matter has been greatly enlarged and complicated in recent years by the insertion of genetical interpretations of plant behavior. It will be impossible to do more than cite a few of the American papers which touch the problem.

HULTÉN's paper on an "Outline of the History of Arctic and Boreal Biota During the Quarternary Period" (Stockholm, 1937) greatly extended many of the ideas already expressed by FERNALD on Glacial and post-Glacial dispersal (see above, "Persistence, etc."). The controversial issues of the "nunatak hypothesis" raised by FERNALD have engendered considerable stimulating discussion, much of which is summarized in papers by Frère MARIE-VICTORIN, "Phytogeographical Problems of Eastern Canada" (Am. Midland Nat. 19: 489–558, 1935); V. C. WYNNE-EDWARDS, "Isolated Arctic-alpine Floras in Eastern North America: a Discussion of Their Glacial and Recent History" (Trans. Roy. Soc. Can. III. 31(5): 1–26, 1937) and "Some Factors in the Isolation of Rare Alpine Plants" (Trans. Roy. Soc. Can. III. 33(5): 35–42, 1939; R. F. GRIGGS, "The Ecology of Rare Plants" (Bull. Torr. Bot. Club 67: 575–94, 1940); and by H. M. RAUP, "Botanical Problems in Boreal America" (Bot. Rev. 7: 147–248, 1941).

Notable advances in our knowledge of the origin and development of the great deciduous forest complex of eastern America have been made in recent years by a number of students. A paper by H. A. GLEASON in 1923 on "The Vegetational History of the Middle West" (Ann. Ass. Am. Geogr. 12: 39–85) has been followed, in the more purely floristic field, by FERNALD's "Specific Segregations and Identities, etc." (see above), E. L. CORE's "Plant Migrations and Vegetational History of the Southern Appalachian Region" (Lilloa 3: 5–29, 1938), E. LUCY BRAUN's papers on the "Affinities of the Flora of the Illinoian Till Plain of Southwestern Ohio" (Rhodora 37: 349–61, 1935) and "Some Relationships of the Flora of the Cumberland Plateau and Cumberland Mountains in Kentucky" (Rhodora 39: 193–208, 1937); and by S. A. CAIN's "Certain Floristic Affinities of the Trees and Shrubs of the Great Smoky Mountains and Vicinity" (Butler Univ. Bot. Stud. 1: 129–150, 1930). Vegetational phases of the problem have been dealt with by Miss BRAUN in a series of papers on the flora of southern Ohio and Kentucky: "Glacial and post-Glacial Migrations Indicated by Relic Colonies of Southern Ohio" (Ecology 9: 284–302, 1928); "The Undifferentiated Deciduous Forest Climax and the Association Segregate" (Ecology 16: 514–19, 1935); "The Differentiation of the Deciduous Forest of Eastern United States" (Ohio Jour. Sci. 41: 235–41, 1941); "The Forests of the Cumberland Mountains" (Ecol. Monogr. 12: 415–47, 1942). The recent status of the eastern prairie-forest boundary problem was summarized in 1935 by E. N. TRANSEAU in "The Prairie Peninsula" (Ecology 16: 423–37).

Some recent papers on the interior plains of the continent are by CLEMENTS and CHANEY on "Environment and Life in the Great Plains" (Carnegie Inst. Wash. Suppl. Publ. 24: 1–54, 1937); FORREST SHREVE on "The Desert Vegetation of North America"

(Bot. Rev. 8: 195–246, 1942); and I. M. JOHNSTON on the "Floristic Significance of Shrubs Common to North and South American Deserts" (Jour. Arn. Arb. 21: 356–63, 1940). The affinities of the vegetation of the northern interior plains have been discussed in papers by RAUP: "Phytogeographic Studies in the Peace and Upper Liard River Regions, Canada" (Contr. Arn. Arb. 6, 1934); "Botanical Investigations in Wood Buffalo Park" (Nat. Mus. Can. Bull. 74, 1935).

There have been but few papers in recent years on the historical plant geography of the North American Cordillera. Problems in the alpine flora of the more northerly areas were dealt with in part by HULTÉN ("History of Arctic Biota, etc."); and the arctic affinities of the alpine flora of the central Rockies were treated somewhat earlier by THEODORE HOLM in "Contributions to the Morphology, Synonymy, and Geographic Distribution of Arctic Plants" (Rept. Can. Arct. Exped., 1913–1918, 5: pt. B, 1–139, 1922), and in "The Vegetation of the Alpine Region of the Rocky Mountains in Colorado" (Mem. Nat. Acad. Sci., Wash. 19^3: 1–43, 1923). G. N. JONES had discussed the phytogeography of the Olympic Mountains of Washington in "A Botanical Survey of the Olympic Peninsula, Washington" (Univ. Wash. Publ. Bot. 5: 1–286, 1936). A recent paper by HERBERT L. MASON, on the "Distribution History and Fossil Record of Ceanothus" (in "Ceanothus", by M. VAN RENSSELAER, pp. 281–303, Publ. Santa Barbara Bot. Gard., 1942), contains a summary of some current ideas on the history of Cordilleran floras.

The genetical approach to problems of migration and area has been adequately summarized very recently by G. L. STEBBINS in "The Genetic Approach to Problems of Rare and Endemic Species" (Madroño 6: 241–58, 1942). These views were put into practice on a regional floristic scale by HULTÉN in his "Arctic and Boreal Biota, etc." (*see above*), and their implications for all students of the boreal American flora have been suggested by RAUP in "Botanical Problems, etc." (*see above*).

Botanical aspects of the theory of continental drift have been almost entirely neglected by American students. A brief paper (in abstract form) by W. H. CAMP on "Continental Displacement and the Origin of American Floras" (Proc. 8th Pan Am. Sci. Cong. 3: 193–4, 1942) is one of the very few that have appeared. WULFF's review of European views on the matter (Chapter X) should, therefore, be of great interest.

HUGH M. RAUP

CONTENTS

Chapter I

HISTORICAL PLANT GEOGRAPHY — SCOPE, RELATION TO ALLIED SCIENCES, METHODS OF INVESTIGATION

Scope and Name of the Science: — Historical plant geography has as its aim the study of the distribution of species of plants now existing and, on the basis of their present and past areas, the elucidation of the origin and history of development of floras, which, in turn, gives us a key to an understanding of the earth's history. In this respect historical geography of plants and animals is a direct continuation of historical geology. The latter science bases its conclusions on a study of fossil organisms, both of animal and plant origin. Consequently, its penetration into a knowledge of the history of our planet carries us no further than the Tertiary or the beginning of the Quaternary Period. From then on the further study of the past fates of the earth passes over to the biologist, who, on the basis of the present distribution of living organisms and of data regarding their past habitats, determines those changes which occurred in that complex combination of diverse factors as a result of the interaction by which the areas of these organisms have been established. By an analysis of these changes he contributes to the task of reconstructing the past aspect of the earth and its history.

The vegetation of our planet is under the constant influence of the most diverse factors facilitating or hampering its development. These factors affecting the earth's vegetation not only functioned in past geological periods, when changes in the configuration and location of the continents, the formation of new mountain systems, the transgression and regression of seas, and changes in climatic conditions called forth changes in the distribution of plants leading to the creation of their present areas; these factors, to which has been added another major one, man's activities, have also continued to function in the present period of the earth's history.

The more recent changes in the earth's vegetation—the disappearance of forests, the formation of deserts, the draining of swamps and changes in their vegetation, the crowding out of some species by others, the destruction or dying out of single species or whole floras and their replacement by cultivated vegetation—which may frequently be traced from historical data likewise constitute one of the chapters in historical plant geography.

This conception of the scope of this branch of botanical geography leads us also to the name by which we designate it, "historical plant geography". The introduction of this term should, we believe, be ascribed in part to STROMEYER but chiefly to SCHOUW, who used it in his "Grundzüge einer allgemeinen Pflanzengeographie" (1822) to designate the branch of science we are discussing. Later ALPHONSE DE CANDOLLE called it "epiontology". ENGLER called it "Entwicklungsgeschichtliche Pflanzengeographie", seeing in the history of the development of floras the chief task of this branch of botanical geography. DIELS and SCHRÖTER gave it the name "genetic plant geography", returning in essence to DE CANDOLLE's term. (For a schematic presentation of the different views on this question see the table below).

TERMS EMPLOYED BY VARIOUS AUTHORS TO DESIGNATE THE THREE MAIN BRANCHES OF PLANT GEOGRAPHY:—

AUTHOR	BRANCHES OF PLANT GEOGRAPHY		
	Floristic	*Ecological*	*Historical*
WILLDENOW (1792)	Geschichte der Pflanzen	Geschichte der Pflanzen	Geschichte der Pflanzen
STROMEYER (1800)	Vegetabilium geographia	Phyto-geographia	Historia vegetabilium geographica
HUMBOLDT (1807)	Geographie der Pflanzen	Geographie der Pflanzen	Geographie der Pflanzen
DE CANDOLLE, AUG. (1820)	Géographie botanique	Géographie botanique	Géographie botanique
SCHOUW (1822)	Pflanzengeographie	Pflanzengeographie	Geschichte der Pflanzen
DE CANDOLLE, ALPHONSE (1855)	—	—	Géographie botanique raisonnée (Epiontologie)
GRISEBACH (1866)	Topographische Geobotanik	Klimatologische Geobotanik	Geologische Geobotanik
DRUDE (1890)	Topographische Geobotanik. Vegetations-physiognomie	Klimatologische Geobotanik	Geologische Geobotanik
ENGLER (1899)	Floristische Pflanzengeographie	Physiologische Pflanzengeographie	Entwicklungsgeschichtliche Pflanzengeographie
DIELS (1908)	Floristische Pflanzengeographie	Oekologische Pflanzengeographie	Genetische Pflanzengeographie
GRAEBNER (1910)	Floristische Pflanzengeographie	Oekologische Pflanzengeographie	Genetische Pflanzengeographie
RÜBEL (1922)	Chorologische Geobotanik	Oekologische Geobotanik	Genetische Geobotanik
SCHRÖTER (1913)	—	—	Genetische Pflanzengeographie oder Epiontologie
HAYEK (1926)	Floristische Pflanzengeographie	Oekologische Pflanzengeographie	Entwicklungsgeschichtliche oder historische Pflanzengeographie

The term "geobotany" was first used in 1866 by GRISEBACH to designate all branches of botanical geography, and it was used in the same sense by DRUDE (1890) and later by RÜBEL (1922–1927).

But in the same year as GRISEBACH, either independently or perhaps under the influence of the "Vegetation der Erde" by this author, there was published in Russia a memoir by RUPRECHT entitled "Geobotanical Studies of the Black-Soil Zone", in which the term "geobotanical" is used not in the broad sense in which it was employed by GRISEBACH but in a narrower sense. By "geobotany" RUPRECHT apparently meant only that part of botanical geography which is concerned with a study of the history of the distribution of species and the development of floras or—as LITVINOV (1895), the well-known specialist on the problems we have under consideration in this book, expressed it—with an elucidation "of the extent to which the age of a land is reflected in the present distribution of plants".

Later, geobotany came to mean only the interrelations between soil and vegetation, *i.e.*, the use of this term was narrowed down to embrace not all but only part of ecological plant geography. And, finally, in recent times geobotany has come to mean the science dealing with plant associations. In view of the great confusion in the use of this term, it seems advisable to adopt some more concrete term.

All the other names proposed likewise have their drawbacks. ENGLER's "Entwicklungsgeschichtliche Pflanzengeographie"—although it comes closest to our concept, stressing precisely as it does problems of the history of the development of floras—is exceedingly cumbrous and difficult to translate into other languages. DIELS' "genetische Pflanzengeographie" embraces only problems of the origin of floras and does not reflect their historical development and present fate; also in sound it very closely resembles an entirely different science, genetics, in which, moreover, at the present time there has developed a special branch known as "genogeography". The term "genetic plant geography" is particularly inacceptable, because it stresses the initial moments in the history of species and floras, entirely ignoring the dynamics of their development and distribution. The task of historical plant geography is to picture the distribution of plants not statically but as a historical process.

It is likewise erroneous to include in this branch of plant geography only the genesis of the areas of species. Historical plant geography aims to elucidate not only the origin and history of the distribution of species but, to no less an extent and even as its chief task, the history of the development of floras, and the genesis of floras may not at all coincide either in place or time of origin with the genesis of many of the species forming these floras.

The term "historical plant geography" has its disadvantages, since "history" is often understood as embracing only those events linked with the period of man's existence, and some botanical geographers (*e.g.*, STROMEYER, 1880, Historia vegetabilium geographica applicata; FLAHAULT, 1907, Phytogéographie historique) have limited this term to designate only those changes in the plant world which have occurred as a result of man's activities.

Nevertheless, we adhere to this last term, "historical plant geography", since we consider that for the given branch of botanical geography, closely linked with historical geology, this is the most suitable designation, being a concept broad enough to embrace all the diverse tasks involved in a branch of knowledge concerned with a study of the development of present-day vegetation in its historical and geographical perspectives.

Relation to Paleobotany: — Studying the present distribution of plants, historical plant geography cannot but investigate their past distribution, since only on the basis of the latter can their present areas be understood. Unfortunately, fossil plants are to be found in very small numbers and often in such a condition that it is impossible to identify them. Nevertheless, even the little that we know about the vegetation of earlier geological periods, particularly of the Tertiary Period, gives us very valuable indications as to the former distribution of genera and species of plants, some of which are still in existence, and this in turn gives a basis for determining the climates and distribution of land areas in those remote times. Consequently, paleobotany is a necessary basis for historical plant geography, but the latter does not cover the same ground as the former, despite the closeness of the aims of research in the two sciences. Paleobotany makes a study not only of the taxonomy, morphology, biology, and geography of fossil plants but also of the history of development of the floras of former geological periods, in this latter respect constituting a science parallel to that of the historical geography of plants now living. *Historical plant geography begins its work at the point where paleobotany leaves off.*

Relation to Phylogenetic Taxonomy: — Phylogenetic taxonomy of plants has as its aim the arrangement of plants (now or formerly existing) in a system based on the degree of their kinship and the history of their development. In connection with this aim modern plant taxonomy should use fully objective methods of determining relationships. At present no one longer doubts that the morphological method of comparison alone cannot give sufficiently trustworthy data as to the relationship of taxonomic units. Consequently, there have been advanced several objective methods of establishing the relationship between species and forms being studied. Among such methods is that introduced by WETTSTEIN (1898) under the name "geographical-morphological method of plant taxonomy". This method has proved to be exceptionally fruitful, and to the present time is widely used in monographs on plant taxonomy. It is founded on the supposition, based on a number of extensive investigations, that there exist close interrelations between species-formation and habitat conditions. Habitat conditions vary not only in relation to time but also in relation to space. Hence, it is quite clear that species arising as a result of adaptation to or the effect of conditions characteristic of a definite area should occupy this area. Consequently, on the basis of the distribution of plants one may draw conclusions as to their origin.

Species may be divided into three main groups according to the conditions and time of their origin. To the first group belong those species which have arisen comparatively recently as a result of adaptation to new habitat conditions, to which they have been subjected either due to their migration beyond the limits of their initial area or due to changes in conditions within part of their area. Such species are, presumably, very closely related to the species from which they arose, inhabit contiguous but not overlapping areas, and are linked by a number of transitional forms not of hybrid origin.

To the second group belong species of more ancient origin. Such species inhabit areas separated from one another but lying within the area of another related species, or they are separated by a region unoccupied due to the dying out of intermediate species, or, in case they already differ sufficiently as regards conditions of habitat, overlapping areas. Such species naturally differ greatly from one another morphologically, and are usually not linked by any transitional forms.

These first two categories of species appeared in northern and central Europe in the post-glacial period in contrast to the third group of species, which existed in the Tertiary Period and were preserved during the Glacial Period either in southern Europe or beyond its boundaries, whence in unaltered form they penetrated Europe proper only after the recession of the glaciers.

From the foregoing it is clear that phylogenetic taxonomy can derive much from the geographical distribution of species in its task of determining their relationships. Historical plant geography, on the other hand, can determine, on the basis of the relationship between species, the history of their origin and migrations, and thus approach its chief task, the elucidation of the history of floras.

Relation to Paleogeography: — Paleogeography is a new branch of geographical science. One of the numerous methods by the aid of which paleogeography approaches a solution of its tasks is the paleo-biogeographical method, founded on a study of the former distribution of living organisms. Paleontological data, due to their chance and inadequate nature, are not capable of solving problems as to the former distribution of plants and animals on the earth's surface, which make it necessary for the paleogeographer to seek in biological data for that which geology fails to give him. Consequently, biogeography constitutes one of the necessary bases for paleogeography.

However, not all branches of the geography of living organisms are equally useful in paleogeographical research. Of exceptional significance for the geography of former periods of the earth's history is the historical geography of plants and animals. The latter, to a considerable degree, have outlived the various changes which have taken place on the earth's surface during their long existence, but these changes cannot but be reflected in the present distribution of organisms. The biologist in studying the latter stumbled on a number of facts, to explain which he took recourse, to some extent without having sufficient geological data, to conclusions and hypothetical propositions as to the former configuration of and connections between the continents, which in many cases subsequently proved to be confirmed by geological investigations and now are generally accepted as established facts.

The study of the present areas of plants and animals, the elucidation of the causes which brought them about, the drawing of conclusions as to the past history of these areas—all these factors in the study of the distribution of organisms, constituting the essence of historical biogeography, make up one of the cornerstones on which paleogeography is based. Paleogeographical data, on the other hand, constitute a basis for conclusions as to the historical geography of plants and animals.

Relation to Paleoclimatology: — Climatic conditions constitute one of the important component factors which, taken together, determine the present distribution of plants. Of exceptional importance for elucidating the history of the former distribution of plants and, consequently, for an understanding of their present areas, is an acquaintance with the climate of past geological periods, when the areas now under investigation were formed.

Conversely, paleoclimatology to a considerable degree bases its work on biogeographical data, not only as regards separate species but also as regards the distribution and character of entire floras. For paleoclimatological research paleogeographical reconstructions are, as Eckardt (1921) has expressed it, "der wichtigste Lebensquell", and, inasmuch as paleogeography is based on biogeographical data, paleoclimatology is also linked with biogeography.

Relation to Historical Geology: — From all the foregoing it is perfectly clear that, if the chief object of historical plant geography is to explain the present distribution of plants on the basis of the history of their past habitats, the geological history of the earth's surface, the history of seas and lands, *i.e.*, historical geology, must constitute an important base for starting points in investigations into historical plant geography. Data for biogeographical conclusions as to the former character of now discontinuous areas of distribution of organisms, data for hypotheses regarding former connections between the continents, without which many factors in this distribution cannot be understood, data as to the presence of seas where there is now land and land where there are now seas, data as to the movement of glaciers and seas — in a word, historical plant geography finds in historical geology all that which reconstructs the history of the earth. Biogeographical conclusions only find definite confirmation, when it is possible to base them on a geological foundation, which historical geology alone can provide. At the same time, biogeography, inasmuch as it is a source of paleogeographical structures, also contributes its bit to the work of historical geology in deciphering the pages of the ancient manuscript of the earth's history.

Methods Employed in a Historico-Geographical Study of Floras: — The history of the development of any given flora may be established on the basis of an accumulation of data, which may be amassed by all the various modes of investigation which the present state of our knowledge places at our disposal. The initial step should be to acquaint ourselves with the geological, paleogeographical, and climatological history of the territory of the flora or floras under study. An analysis of present-day floras, based on a study of the areas of their component species, should next be made both by the direct method of a study of paleobotanical data and by a number of indirect methods, such as the phylogenetic, botanico-geographical, ecologico-phytocoenological, and biotic. By the last-mentioned we mean a comparison of the distribution of plants with that of animals or of the distribution of plant hosts with that of their parasites.

Historico-geographical conclusions should be based, first of all, on paleobotanical data, despite the casual and fragmentary nature of the latter. The importance of such data should not be underestimated, since they constitute the only direct evidence available for establishing the history of a flora.

By the phylogenetic method we gain a knowledge of the species composing a given flora and of their areas. Deeper investigations along this line lead to monographic studies of certain groups of species most characteristic of selected genera. On the basis of such studies the phylogenetic direction of development of these genera is established, which, combined with data from a study of geographical distribution, points to their centers of origin and to the history of their dispersal from these centers. Combining the results obtained from such a study of genera representing the various floristic groups, we may draw conclusions as to the trend of development of the entire flora under investigation.

The establishment of phylogenetic links and relationships between the species composing a given genus is achieved by all possible scientific methods, of which the chief is the morphological method. The sequence of changes involving increased complexity of structure of species (morphological, cytological, anatomical, biochemical, etc.) may coincide with changes in the geographical location of individual species or whole sections of a genus.

The trend of evolution of a genus from primitive representatives to those of more complex structure, from relic forms showing but slight variation to progressive forms, as reflected in their geographical distribution, gives indications as to the probable initial center of development of a genus, its subsequent geographical migration and the routes of the latter, the formation of secondary centers of diversity, the appearance during the course of such migration of vicarious species, etc. Such a taxonomic study of the genera composing a given flora constitutes a firm foundation for a simultaneous or subsequent study by the botanico-geographical method. If the geographical method has proved to be exceptionally valuable in a taxonomic study of plants, the taxonomic method is, conversely, of no less importance in a historico-geographical study of floras. In every modern botanical monograph may be found numerous examples illustrating the possibility of such use of taxonomic data as an aid to historical plant geography. We shall cite only a few instances of the founding of historico-geographical conclusions on the basis of a phylogenetic study.

One such case is that of the establishment of the chief features of the history of development of the flora of Mongolia and China by V. L. Komarov (1908) on the basis of a study of five genera chosen on account of the nature of their geographical distribution. A monographic study of these genera made it possible to determine the centers of their origin and the routes of their subsequent migration, which, taken together, indicated the trend of development of the entire flora of Mongolia and China. As a second example we may take the phylogenetic study of the genera of the family *Sapotaceae* made by Lam (1935), who found that there is a decrease in the number of sepals in the flowers of the more highly organized species and also that there are changes in a number of other morphological characters specific for the four genera of this family. The trend of geographical distribution of these species — from the Malay Archipelago toward the west in the direction of India and toward the east in the direction of New Guinea —indicates that the Malay Archipelago should be considered the initial center of development of the *Sapotaceae*.

The botanico-geographical method is based on a study of the areas of the species of a given flora, not only within the limits of the latter but, which is most important, in all their entirety. As a result of such an investigation, the centers of concentration of species may be established, which point to the centers (primary or secondary) of development of genera. Moreover, by establishing the character of an area, it is possible to determine of what ecological, geographical, and historical elements a given flora is composed. Combining these data with those obtained from a taxonomic study makes it possible to subdivide the established geographical elements—according to the centers of origin of the areas of the various species—into a number of groups reflecting the genesis and process of development of the flora under investigation, the degree of its autochthonism, *i.e.*, the extent to which the species composing it have originated and developed within the territory occupied by the given flora, the extent to which such species have migrated from neighboring floras, and the time and direction of these immigrations. A comparison of the flora under investigation, from the indicated points of view, with analogous floras of other countries, makes it possible to elucidate the mutual relations of these floras, while the combined study of the floras of various countries and regions gives a picture of the history of development of the flora of the entire globe.

Lastly, changes in vegetation during the most recent period of the earth's history, when it has been affected by man's activities, are studied both on the basis of data regarding the interrelations of species in plant associations and on the basis of historical documents and material. A study of the ecology of species in different habitats within their areas may elucidate the biological peculiarities of the initial types and thereby indicate the direction of dispersal of a given genus or species.

In the case of all the above-mentioned indirect methods it is of utmost importance to study not separate species but a geographical series of species that replace one another throughout the area of the genus to which they belong. The difficulties confronting us in this field of science, only the initial steps in whose development have so far been made, are very great, but there are no grounds for regarding them as insuperable.

References:

DIELS, L., 1921: Die Methoden der Phytographie und der Systematik der Pflanzen (Abderhalden's Handb. d. biol. Arbeitsmethoden, Vol. 11, Part I, No. 2).

ECKARDT, W., 1921: Die Paläoklimatologie, ihre Methoden und ihre Anwendung auf die Paläobiologie (Abderhalden's Handb. d. biol. Arbeitsmethoden, Vol. 10, No. 3).

GAMS, H., 1926: Pflanzengeographie, Paläogeographie und Genetik (Petermann's Geog. Mitteilungen).

KARSTEN, G., 1922: Methoden der Pflanzengeographie (Abderhalden's Handb. d. biol. Arbeitsmethoden, Vol. 11, Part I, No. 3).

KOMAROV, V. L., 1908: Introduction to the floras of China and Mongolia (In Russian; Acta Horti Petropolitani, Vol. 29).

LAM, H. J., 1935: Phylogeny of single features (Garden's Bull. Straits Settlements, Vol. 9, Part 1).

LITVINOV, D. I., 1891: Geobotanical Notes on the Flora of European Russia (In Russian; St. Petersburg).

RÜBEL, E., 1927: Ecology, plant geography and geobotany, their history and aim (Bot. Gazette, Vol. 84, No. 4).

RUPRECHT, F., 1866: Geobotanical Studies of the Black-Soil Zone (In Russian; Suppl. to Vol. 10, Notes Imp. Acad. Sci., No. 6, St. Petersburg).

SCHWARZ, O., 1938: Phytochorologie als Wissenschaft, am Beispiele der vorderasiatischen Flora (Fedde's Repert. Spec. Nov., Beihefte, Vol. 100).

WETTSTEIN, R., 1898: Grundzüge der geographisch-morphologischen Methode der Pflanzensystematik (Jena).

Chapter II

HISTORY OF THE SCIENCE

The first definitely expressed ideas on regularities in the distribution of plants, which we may regard as the beginning of historical plant geography as a science, we find in the work of WILLDENOW, "Grundriss der Kräuterkunde", which first appeared in 1792. In the seventh section of this work, entitled "Geschichte der Pflanzen", we read:—

"By history of plants is meant the influence of climate on vegetation, *the changes which plants have probably undergone as a result of the revolutions which have taken place on our globe, their distribution over the earth's surface, their migrations*, and, lastly, the provisions nature has made for their preservation" (p. 418).*

The character of the present distribution of plants caused these questions to arise in WILLDENOW's mind: Did not the seas occupy more space in former times than now? Was not the globe entirely covered with water, from which projected only the mountain peaks, which in those times were the only habitat of plants? As the seas dried up and the area of dry land increased, the plants began gradually to disperse from these initial habitats. Later hurricanes, earthquakes, and volcanoes again destroyed plant life over large areas. This is evidenced by plants whose distribution is restricted to small, widely separated localities. "Lands now separated by oceans may, in former epochs, have been united . . . Thus, the northern part of America may have been connected with Europe, New Netherlands ** with the foothills of the Cape of Good Hope".

In addition to these purely historical causes, WILLDENOW points out a number of extant factors affecting the distribution of plants. Among such factors he mentions the various adaptations for the dispersal of fruits and seeds by the aid of animals, wind, river, and sea currents, and the scattering of seeds by dehiscent fruits. A no less important factor in the distribution of plants is man. He also points out the similarity between aquatic plants and plants growing on mountain peaks, and in conclusion he discusses the origin of various floras. Hence, in this brief treatise we already find elements of modern historical plant geography.

The next work in order of time, which is clearly concerned not with the geography of plants in general but precisely with problems of historical geography, is that of STROMEYER, entitled "Commentatio inauguralis sistens historiae vegetabilium geographicae specimen". In this work a clear distinction is made between general plant geography, "vegetabilium geographia" (phytogeography), and historical plant geography, "historia vegetabilium geographica", of which he makes a further subdivision, "historia vegetabilium geographica appli-

* Cited from the second edition, Vienna, 1798; italics ours.
** Formerly used to designate Australia.

cata", which treats of the distribution of plants as linked with the history of the settlements and migrations of men and animals.

Nevertheless, it is ALEXANDER VON HUMBOLDT who should be regarded as the founder of plant geography, including also problems of the origin of floras, since it was he who, in his "Essai sur la Géographie des Plantes" ("Ideen zu einer Geographie der Pflanzen" in the German edition), appearing in 1807, established this new science and gave it its present name. In this work HUMBOLDT sets forth his ideas, which embrace the bases of modern botanical geography, including problems with which historical plant geography is concerned:

> "In order to come to a decision as to the existence in ancient times of a connection between neighboring continents, geology bases itself on the analogous structure of coastlines, on the similarity of animals inhabiting them, and on ocean soundings. Plant geography furnishes most important material for this kind of research. It can, up to a certain point, determine the islands which, at one time united, have become separated from one another; it finds that the separation of Africa and South America occurred before the development of living organisms. It is again this science that shows which plants are common to both eastern Asia and the coastlands of Mexico and California, and whether there are some which grow in all zones and at all altitudes. It is by the aid of plant geography that we can go back with some certainty to the initial physical state of the globe. It is this science which can decide whether, after the recession of the waters to whose abundance and movements the calcareous rocks attest, the entire surface of the earth was covered simultaneously with diverse plants, or whether, according to the ancient myths of various peoples, the globe, having regained its repose, first produced plants only in a single region, from which the sea currents carried them progressively, during the course of centuries, into the most distant zones" (pp. 19–20)*.

No less definitely is put the question as to the importance of a study of the past and present distribution of organisms: "In order to solve the great problem as to the migration of plants, plant geography descends into the bowels of the earth; there it consults the ancient monuments which nature has left in the form of petrifactions in the fossil wood and coal beds which constitute the burial-places of the first vegetation of our planet" (p. 22). The finding in temperate zones of the remains of plants and animals of warmer climes puts to the fore the question as to the former climatic conditions in the given localities.

A still more detailed understanding of the tasks of botanical geography we find in the works of AUGUSTIN P. DE CANDOLLE, who in his "Essai Elémentaire de Géographie Botanique" (1820), and also in other works, clearly distinguishes between the "habitation" and the "station" of a plant, meaning by the first the distribution of a plant over the earth's surface and by the second its habitat conditions as a whole. Thus, the section of this work entitled "Des habitations" constitutes in considerable measure historical plant geography as he understood it. In this chapter he does not limit himself to establishing the habitats of plants but attempts to determine the causes underlying this or that kind of distribution. Factors such as seas, deserts, mountain chains, swamps, forests, and variations in altitude constitute obstacles to dispersal. Plants are endowed in different degree with the ability to overcome these obstacles, and are dependent in great measure on passive factors as an aid in surmounting them. Such factors

---

* Cited from the French edition.

include sea and river currents, atmospheric currents, animals, and man. If these four means of seed dispersal are kept in mind, "one will find, I believe, that they are fully adequate to explain the finding of a small number of plants common to different continents . . . Their combined action—slow, steady, imperceptible—constantly tends to disperse plants in all directions, and these plants become naturalized where they find conditions favorable for their existence" (p. 410).

After the appearance of the above-mentioned works it was possible to publish the principles of the new science, the first attempt in this direction being SCHOUW's "Grundzüge einer allgemeinen Pflanzengeographie", issued in Danish in 1822 and translated into German in 1823. To the development of historical plant geography, however, this work contributed nothing, since its author makes a sharp distinction between "plant geography" proper and the "history of plants", to which latter he refers all problems with which historical plant geography is concerned. Thus, WILLDENOW's term "history of plants", embracing all botanical geography, is narrowed down by SCHOUW, who includes in it only problems of the history and genesis of species and floras, which even earlier, as we have seen, were touched upon by STROMEYER. SCHOUW regards the "history of plants" as an independent science. "The history of plants . . . does not constitute a part of physical geography, since it is not a descriptive science, but rather a part of the history of the earth, inasmuch as the latter treats not only of inorganic but also of organic bodies" (p. 10).

In another similar work by MEYEN, "Grundriss der Pflanzengeographie", which appeared thirteen years after SCHOUW's book, we also find almost nothing concerning historical plant geography, with the exception of a few pages devoted to problems of areas of distribution and their determination.

In contrast to these works, a considerable contribution to historical plant geography was made by UNGER's "Versuch einer Geschichte der Pflanzenwelt", which appeared in 1852. This book, speaking in present-day terminology, may be considered a treatise on paleobotany, embracing, however, relations between fossil and existing floras. The introduction to this book, in which the author discusses the distribution of plants, constitutes a notable contribution to historical plant geography.

Although the development of this new branch of knowledge, as we have seen, had its expression in a number of works inspired in considerable measure by VON HUMBOLDT, the credit for elaborating the latter's ideas and for making the first synthesis of the new science of historical plant geography belongs to ALPHONSE DE CANDOLLE. In his chief work, published in 1855 and entitled "Géographie botanique raisonnée ou exposition des faits principaux et des lois concernant la distribution géographique des plantes de l'époque actuelle", DE CANDOLLE, as is clear from the title itself, undertook the task of elucidating the laws of the distribution of plants. His understanding of what this task involves is set forth in the preface to this work (pp. xi-xii): —

"Plants have a habitat conforming to the climate only under certain circumstances, only in certain countries; there is not a single botanist but who knows that a species can ordinarily live and reproduce itself far from its native home and that no country contains all the species which can live there without man's protection. These are facts, one has been wont to say. Yes, there are facts, but why these facts? What are their causes—possible, probable, or certain? . . . And if, on the basis of the present distribution of species and of a knowledge of climatic conditions, one could come to understand their original distribution, would that not be a splendid achievement of science?

"These important questions were for a long time a mystery to me. But I was far from inclined to fight shy of them, as some authors have done. On the contrary, these questions attracted me, tormented me. I perceived neither solutions nor ways to arrive at solutions.

"Fortunately, the advances in geology have shed on natural science a new light. This light, no doubt, began as a feeble glimmer, but one penetrating everywhere. Now it is becoming larger; it shows us extensive vistas, entirely new. We can try to go back in the chain of times to the origin of the vegetable and animal kingdoms. We have come to the conclusion that the living organisms of our epoch have passed through diverse climatic conditions and past geographical conditions no less varied. Thus, when the present distribution of species seems odd to us, when it does not conform to modern climatic conditions, it is probable that this is the effect of former geological and physical conditions. We see here only the result of a different order of things, which in its turn was the consequence of still different preceding conditions.

"From this new point of view botanical geography ceases to be a simple accumulation of facts. It occupies, on the contrary, an imposing position in the center of the sciences. Its principal aim should be *to show what, in the present distribution of plants, may be explained by present climatic conditions and what is a consequence of former conditions.**

"By assigning it such a high aim, botanical geography competes with the history of fossil organisms (paleontology) and with geology proper in research on one of the greatest problems of natural science or, rather, of science in general and of all philosophy. This problem is that of the succession of organisms on the globe."

As an approach to the assigned task, DE CANDOLLE, in the early chapters of his book, discusses the effect of external factors—temperature, light, and humidity—on the distribution of plants and also takes up the different types of distribution. The data assembled lead him to the problem of areas of distribution, their character, the changes which they undergo, and the establishment of a number of regularities in the formation of areas.

This famous work of DE CANDOLLE is exceptionally rich in content, and we shall have need to refer to it repeatedly. But here, in our historical review, we must limit ourselves to this very brief account.

Somewhat earlier (in 1846) the remarkable work of FORBES, "On the connexion between the distribution of the existing fauna and flora of the British Isles and the geological changes which have affected their area, especially during the epoch of the Northern Drift", opened new vistas for the understanding of the present geographical distribution of plants. As a basis for this work was LYELL's important book, "Principles of Geology", which first appeared in 1832 and exerted an immense influence on the development of biological views, including problems connected with the geographical distribution of organisms. Such influence was furthered not only by the entirely new point of view of the author on the geological past of the earth, but also by the fact that LYELL considered it necessary to include in his work also "phenomena relating to the organic world, which have an equal claim on our attention, if we desire to obtain possession of all the preparatory knowledge respecting the existing course of nature, which may be available in the interpretation of geological monuments" (p. 566). Conse-

* Italics in the original.

quently, among chapters devoted to problems concerning the genesis of species, we find chapters on the geographical distribution and migrations of species, insular flora and fauna, and the extinction of species.

FORBES, as a starting point for his subsequent exposition, assumes the existence of "specific centres", *i.e.*, of "certain geographical points from which the individuals of each species, originating from a single progenitor or from two, began their geographical distribution." To substantiate this view, FORBES points out the following three facts: "(1) Species of opposite hemispheres placed under similar conditions are representative and not identical. (2) Species occupying similar conditions in geological formations far apart, and which conditions are not met with in the intermediate formations, are representative and not identical. (3) Wherever a given assemblage of conditions, to which, and to which only, certain species are adapted, are continuous—whether geographically or geologically—identical species range throughout" (p. 336). Examining from this point of view the flora of the British Isles, FORBES comes to the conclusion that the interrelations between the various elements of their flora may be explained only by migration of its species prior to the separation of the islands from the continent, of which they were formerly a part.

The work of FORBES, as well as the new trend in the understanding of the geological past of the earth initiated by LYELL, opened a new page in the study of the historical geography of plants. One of the first to develop these ideas further was JOSEPH DALTON HOOKER. Thanks to his intimate knowledge of the vegetation of almost the entire globe, acquired by him during his numerous travels and also by a study of the floras of many lands, HOOKER possessed a breadth of vision unattained by investigators before his time. Consequently, in his works we find for the first time a transition from the study of the distribution of separate units in the plant kingdom to an explanation of the origin and development of entire floras.

One of HOOKER's investigations in the field interesting us is that on the vegetation of the Galapagos Archipelago, reported on by him at a session of the Linnean Society in London in 1846 and published in 1851, based on a study of herbarium specimens collected on these islands by DARWIN. The origin of this flora, in HOOKER's opinion, is to be explained as the result of the transport of its component species, particularly the non-endemic species, from the American continent to these islands by ocean currents, wind, birds, and, to a small extent, by man, and their modification under the influence of isolation.

A few years later, in Part II of his "Botany of the Antarctic Voyage", dealing with the flora of New Zealand and published in 1853, HOOKER devotes to problems of the distribution of species a very valuable "Introductory Essay". He points out that no other branch of botany requires for its understanding such an intimate knowledge of plants and of relationships between species as does that concerned with a study of the geographical distribution of plants. Basing himself on the works of LYELL and FORBES and on the premise that one and the same species can have arisen only in one place on the globe, HOOKER draws the conclusion that the plants now distributed on the various islands of the Antarctic at one time formed part of a single flora, occupying a continent larger than that now found in the Antarctic Ocean.

Part III of this same work, devoted to the flora of Australia and Tasmania, appeared in 1860, after the publication of DARWIN's "Origin of Species", which had a great influence on HOOKER but did not advance any views seriously contradictory to his own. In this Part III we find two sections of particular interest: 3. "On the general phenomena of distribution in area" and 4. "On the general phenomena of the distribution of plants in time". In the former the author draws the conclusion that many phenomena in the present distribution of plants cannot be explained by existing factors and that to understand them it is necessary to study past changes in climate and in the distribution of dry land. In Section 4 HOOKER examines the paleobotanical data available at the time of the publication of his work and comes to the conclusion that the problem of the distribution of plants is exceedingly complex. He advances the proposition that changes in the surface of our planet—lands being replaced by seas and valleys by mountain chains—take place in a relatively short interval of time, as compared to the age-long existence of some genera and even species of plants.

In 1855 HOOKER began publication of his "Flora Indica". In the first and only volume of this "Flora" (it was later replaced by his "Flora of British India") there is an "Introductory Essay" of exceptional importance for an understanding of the history of development not only of the flora of India but also of modern tropical flora in general. In 1862 there appeared another important work, "Outlines of the distribution of arctic plants", giving an analysis of arctic flora and data for the establishment of its origin. Of great interest also is HOOKER's "Lecture on Insular Floras"—delivered in 1866 before the British Association for the Advancement of Science—in which he again returns to the problem of the origin of insular floras, their interrelations, and their relations to continental floras.

Reviewing HOOKER's work, we see that he put a number of entirely new problems, concerned not so much with regularities in the distribution of separate species as with the question of the origin of whole floras, determined on the basis of an analysis of the areas of distribution of the component species. This new direction of investigation proved in the highest degree fruitful and constituted a great impetus to work in this field.

From the foregoing it is evident that by the middle of the nineteenth century the question as to the geographical distribution of organisms was for many investigators, such as LYELL, FORBES, HOOKER, and others, in its basic features entirely clear. Nevertheless, the doctrine as to the immutability of species and their existence as a result of separate acts of creation was still firmly rooted, due chiefly to its connection with religious views, and shackled the minds even of outstanding thinkers.

In 1859 there appeared DARWIN's "Origin of Species". The revolution which this book produced not only in biology but in all man's thinking naturally affected the development of our branch of science as well. In his investigations of the evolution of organisms and its causes DARWIN could not but treat problems of their geographical distribution. Chapters XII and XIII of his book are devoted to these problems, and they, consequently, constitute a most valuable contribution to historical plant geography.

These chapters DARWIN introduces with the following statement: "In considering the distribution of organic beings over the face of the globe, the first great fact which strikes us is that neither the similarity nor the dissimilarity of the inhabitants of various regions can be wholly accounted for by climatal and other physical conditions" (p. 493). Having cited a number of examples of the distribution of plants and animals in various countries testifying to the connection between portions of areas of separate species and entire floras and faunas, DARWIN writes: "We see in these facts some deep organic bond, throughout space and time, over the same areas of land and water, independently of physical conditions. . . . The bond is simply inheritance, that cause which alone, as far as we positively know, produces organisms quite like each other, or, as we see in the case of varieties, nearly alike" (p. 497).

DARWIN bases his views on the geographical distribution of organisms on the assumption that each species was first produced in one area alone, subsequently migrating from that area. He concludes: "The endurance of each species and group of species is continuous in time; . . . so in space, it certainly is the general rule that the area inhabited by a single species, or by a group of species, is continuous, and the exceptions, which are not rare, may . . . be accounted for by former migrations under different circumstances, or through occasional means of transport, or by the species having become extinct in the intermediate tracts" (p. 564).

Hence, the finding of the same species on the British Isles and in Europe is fully understandable, since these lands doubtless were at one time united; likewise understandable is the absence of European mammals in Australia and South America, despite similar habitat conditions, confirmed by the naturalization in these countries of many European plants and animals. The existence of identical species of plants separated by great distances is explained by their possession of means of dispersal enabling them to overcome these distances.

What are these means of dispersal? Having discussed the views of LYELL and FORBES on the former connection between Atlantic islands and Europe and between the latter and America, DARWIN writes: "Other authors have thus hypothetically bridged over every ocean, and united almost every island with some mainland. If indeed the arguments used by FORBES are to be trusted, it must be admitted that scarcely a single island exists which has not recently been united to some continent. This view cuts the Gordian knot of the dispersal of the same species to the most distant points, and removes many a difficulty; but to the best of my judgment we are not authorised in admitting such enormous geographical changes within the period of existing species" (p. 505).

Allowing for the possibility of "great oscillations in the level of the land or sea" and also of the "existence of many islands, now buried beneath the sea, which may have served as halting-places for plants and for many animals during their migration", DARWIN nevertheless considers impossible "such prodigious geographical revolutions within the recent period, as are necessary on the view advanced by FORBES and admitted by his followers. The nature and relative proportions of the inhabitants of oceanic islands are likewise opposed to the belief of their former continuity with continents. Nor does the almost universally volcanic composition of such islands favour the admission that they are the wrecks of sunken continents;—if they had originally existed as continental mountain ranges, some at least of the islands would have been formed, like other mountain summits, of granite, metamorphic schists, old fossiliferous and other rocks, instead of consisting of mere piles of volcanic matter" (pp. 505–6).

Starting from these premises, DARWIN proceeds, with the painstaking care so characteristic of him, to assemble facts on the distribution of plants and to test some of them experimentally. He presents data of experiments on the resistance of seeds to the action of sea-water and on the length of time fruits or parts of plants with fruits may float, and he recounts his observations on the transport of seeds on drift timber and icebergs and also by birds, on the distribution of fresh-water plants and animals, and on the inhabitants of islands and their relation to those of the nearest mainland.

Nevertheless, despite the existence of such means of seed dispersal, he concludes: "The floras of distant continents would not by such means become mingled; but would remain as distinct as they now are." In spite of the occasional cases of seeds being transported across the ocean, "how small would be the chance of a seed falling on favorable soil, and coming to maturity! . . . Out of a hundred kinds of seeds or animals transported to an island, even if far less well-stocked than Britain, perhaps not more than one would be so well fitted to its new home as to become naturalised. But this is no valid argument against what would be effected by occasional means of transport, during the long lapse of geological time, whilst the island was being upheaved, and before it had become fully stocked with inhabitants. On almost bare land, with few or no destructive insects or birds living there, nearly every seed which chanced to arrive, if fitted for the climate, would germinate and survive" (pp. 514–5).

Subsequent authors, however, gave to such chance factors primary significance, resorting to them in all cases of an otherwise inexplicable station separated from the main area, not realizing that in reality they were merely substituting such chance transport for the dogma of multiple centers of species creation without supplying any factual proof. From this impasse only in our own times has there been found a way out, in the form of WEGENER's hypothesis of continental drift, according to which, assuming the permanence not of the separate continents and oceans but of the areas occupied by them, we are at the same time able to consider that they were formerly connected, thus solving a number of puzzling problems of biogeography.

Not less important and instructive for subsequent investigators were DARWIN's views on the dispersal of plants during the Ice Age or Glacial Period, on the presence of identical species on isolated mountain summits, on the former existence of land-bridges between the Old and New World, over which in the Tertiary Period, thanks to the warmer climate, there took place an exchange of plants, and on the similarity of the floras of the southern shores of America, Australia, and New Zealand, all of which floras give evidence of having had at one time connection with the flora of the now ice-covered Antarctic continent.

A further contribution to the development of the science interesting us was made by a contemporary of DARWIN, ALFRED RUSSEL WALLACE. The latter may with full right be considered the founder of the science of the geographical distribution of animals. His investigations, concerned only in an inconsiderable part with the plant kingdom, have, nevertheless, because of the general propositions advanced by him, significance in the history of the study of the distribution of plants. For historical plant geography of particular value is his work "Island Life", dealing with the origin of insular faunas and floras. The basic theory of WALLACE, set forth in detail in this work, is summarized in his concluding chapter as follows:

"The distribution of the various species and groups of living things over the earth's surface and their aggregation in definite assemblages in certain areas is the direct result and outcome of a complex set of causes, which may be grouped as 'biological' and 'physical'. The biological causes are mainly of two kinds—firstly, the constant tendency of all organisms to increase in numbers and to occupy a wider area, and their various powers of dispersion and migration through which, when unchecked, they are enabled to spread widely over the globe; and secondly, those laws of evolution and extinction which determine the manner in which groups of organisms arise and grow, reach their maximum, and then dwindle away, often breaking up into separate portions which long survive in very remote regions." Among physical causes WALLACE mentions: "geographical changes, which at one time isolate a whole fauna and flora, at another time lead to their dispersal and intermixture with adjacent faunas and floras" and "changes of climate which have occurred in various parts of the earth,—because such changes are among the most powerful agents in causing the dispersal and extinction of plants and animals" (pp. 531–2).

One of the critics of natural selection, MORITZ WAGNER, in his analysis of DARWIN's theory dwelt on the above-mentioned Chapters XII and XIII of "The Origin of Species". In his chief work, "Die Darwinische Theorie und das Migrationsgesetz der Organismen", which appeared in 1868, and in a number of articles, WAGNER advanced his "law of migrations", which, in his opinion, did not refute but supplemented DARWIN's views. This law of migrations is based on the following propositions: The competition of organisms and the struggle for existence give an impulse to plants and animals to extend the area of their distribution. The new habitat conditions in which the migrants find themselves induce marked variations in their characters. If the obstacles which a species has overcome during its migrations are inconsiderable, so that between the new varieties and the parental forms there remains a close bond, then these varieties by intercrossing very quickly disappear, being blended with the initial forms. In the opposite case they remain distinct and are converted into new geographical races or new species. Conditions in former geological periods—such as changes in the relative distribution of seas and dry land, the breaking off of islands from the mainland, and frequent volcanic eruptions—facilitated to a considerably greater degree the formation of new species than do the conditions now existing.

This law of species formation took shape in WAGNER's mind as a result of observing the existence in nature of very closely related species, often separated by comparatively small barriers, such as rivers, mountains, gulfs, etc., or growing on isolated mountain peaks. These so-called "vicarious" species arose as a result of the chance transport of the basic form across the indicated barriers, where, under the influence of new conditions and of isolation, it became modified and was transformed into a new race. The law of migrations, which provides a key to the understanding of the areas of distribution of vicarious species and of breaks in such areas, is of great significance for historical plant geography.

An outstanding event in the field of study of botanical geography, and particularly of its historical phase, was the publication in 1878–1882 of ENGLER's "Versuch einer Entwicklungsgeschichte der Pflanzenwelt, insbesondere der Florengebiete, seit der Tertiärperiode". The very fact of such a work being published indicates the great progress that historical plant geography had made by that time. For, as we have already pointed out, at the time of DE CANDOLLE the mere putting of the question as to the causes and regularities of the observed distribution of plants constituted an untouched theme, avoided by most authors on account of the difficulties involved. Moreover, in ENGLER's work we find the principles and data of historical plant geography applied not only for an explanation of the distribution of separate units of the plant kingdom but also for an explanation of the development and interrelations of the floras of the entire globe—a problem, as we have seen, first tackled by HOOKER.

Later investigations by ENGLER himself and by his followers and students contributed much that was new, necessitating certain changes in ENGLER's views on the development of floras. One of ENGLER's services was the utilization of data on the geographical distribution of genera and species of plants for his conclusions on the history of the development of floras. Such data he obtained from monographs on selected genera or species, having as their aim to check kinships by a study of areas of distribution. ALPHONSE DE CANDOLLE in his "Géographie botanique raisonnée" (1855) was the first to note the need of utilizing geographic distribution in making monographic studies of genera and families. Five years later, in 1860, STUR presumably was the first to apply this principle in his monograph on the genus *Astrantia*. This same year BORSHCHOV (BORSZOW) published his monograph on the Aral-Caspian species of the genus *Calligonum*, in which he gives an analysis of their geographical distribution.

Somewhat later, in 1869, KERNER published a remarkable work on the dependence of a plant on its geographical distribution in connection with the effect of soil and climatic conditions, based on a monographic study of species of the group *Tubocytisus* of the genus *Cytisus*. During the succeeding years this geographical principle was lent serious support in a number of excellent monographs (on *Acantholimon*, 1872; *Labiatae persicae*, 1873; *Chenopodiaceae*, 1880), published by BUNGE in the transactions (Mémoires) of the Academy of Sciences of St. Petersburg. In 1872 there also appeared ENGLER's monograph on the genus *Saxifraga*, in the title of which it is noted that special attention is given to the geographical phase. A number of very important propositions regarding the geographical distribution of plants were advanced by KERNER in many of his later works.

This new trend in plant taxonomy, which during the ensuing years was extensively developed, was of very great service to historical plant geography, since, on the basis of such thoroughly investigated material, work in the field of the study of the interrelations and development of floras acquired a firm foundation.

Around ENGLER there developed a whole school of botanical geographers, who directed their work on historical plant geography along the lines indicated by him. A great memorial to this school is the collection of botanico-geographical monographs edited by ENGLER and DRUDE under the general title "Vegetation der Erde".

These investigations were provided with an even firmer foundation and developed more rapidly after the publication, in 1898, of the work we have already mentioned by WETTSTEIN on the botanico-geographical method in plant taxonomy and its further elaboration by V. L. KOMAROV in the introduction to this "Flora of Manchuria" (1901).

The number of works in our field of science from this time on increases so rapidly that it is no longer possible to review them as fully as we have up to this point, and we shall now limit ourselves to an enumeration of the chief problems and investigations that have played an important rôle in the development of historical plant geography.

Among such investigations are those of NATHORST and ANDERSSON on fossil plants of the Ice Age. The first irrefutable proof of the former extension of arctic flora considerably further south than its present southern limits was established in 1842 by STEENSTRUP, who found in Danish peat bogs a number of fossil remains of plants giving undoubted indications as to the past history of the vegetation of Denmark. However, the fact of the finding of fossil flora of the Ice Age received full recognition only later, when in 1870 NATHORST found representatives of this flora in eight different localities in southern Sweden. By his detailed studies and those of other investigators the presence of glacial flora was established in very many localities not only in Sweden but also in Norway, Denmark, the Baltic Region, Germany, England, Scotland, Switzerland, Hungary, France, etc. These findings gave entirely unexpected indications as to the migrations of species during and immediately following the Ice Age. Moreover, these investigations showed that the fossil remnants of Ice Age flora found in a single locality were not identical but showed variations depending on the age of the deposits in which they were embedded; the more recent a deposit the more closely did the species composition of its fossil flora resemble that of the flora of the locality. The numerous investigations along this line which followed directly after those of

Nathorst enabled Andersson in 1897 to outline the history of the vegetation of Sweden and of northern Europe in general.

Closely linked with the problem of the Arctic flora is that of the origin and history of development of the Alpine flora, of the finding in the Alps of species of Arctic flora and the finding of Alpine plants far to the north. On these problems many papers have appeared by various investigators. First among these papers was that by Christ (1867).

Another important question of historical plant geography, the interrelations between the floras of North America and eastern Asia, was raised as early as 1846 by Asa Gray and discussed by him in greater detail in a number of subsequent papers. The accumulation of paleobotanical data pointing to the fact that a number of genera, such as *Liquidambar, Sassafras, Aralia, Magnolia, Liriodendron, Taxodium, Sequoia,* etc., represented in our day in North America by various species but absent in Europe, had in the Tertiary Period an extensive distribution, being found as far north as Greenland, gave grounds for Gray to conclude that the flora of North America in the Tertiary Period was closely allied and in many respects identical to the flora then found in Greenland, Spitsbergen, and northern Europe.

A second very important service of Gray was that he pointed out the existence of an undoubted connection between the flora of the Atlantic states of North America and that of northeastern Asia. This remarkable fact, confirmed by an extensive list of species, Gray explains as follows: The flora, of which these species constitute remnants, was distributed in the Miocene in what is now the arctic and subarctic zone. As the climate became colder and the glaciers advanced, this flora retreated toward the south and survived until the present time in those regions that preserved climatic conditions approximating those of former times, as occurred in eastern Asia and Japan, on the one hand, and on the western and eastern shores of North America, on the other. These data indicate that during the Tertiary Period there was a direct connection and interchange of forms between Asia and America, a land-bridge presumably existing at that time where the Bering Strait now lies. In 1859 Maximovicz presented data showing that the flora of eastern Asia had preserved much of its Tertiary character.

Among other investigations of importance for historical plant geography we may mention those concerned with the problem of peat bog vegetation, which have made it possible not only to establish the succession of floras in the post-glacial period but also to determine the past areas of existing species of woody plants. These data, gleaned from a study of microscopic plant remains in peat-deposit profiles, acquired a firmer foundation and developed more rapidly after Weber (1896) and Lagerheim (1905) had proposed a special microscopic method of investigating the plant remains in peat bogs. This method is based on a study of pollen found by taking a number of successive samples at different depths in peat deposits. Thanks to the taxonomic specificity of the structure of pollen, it is relatively easy to determine to which genus and sometimes to which species it belongs. On the basis of such investigations it is possible to judge as to the species composition of a past forest, as to the quantitative interrelations of the species forming it, and as to the changes in species composition of the woody vegetation during the various stages in the post-glacial period. These data constitute exceptionally valuable material for an understanding of the areas of present-day vegetation and of the changes which these areas have undergone during the Quaternary Period. The investigations now under way of pollen and spores preserved in loess (SUKACHEV, 1937) and more ancient deposits, *e.g.*, in Tertiary, Jurassic, and Carboniferous coals, will, in all probability, disclose much that is new as regards the distribution of genera of present and past floras.

By the end of the nineteenth century botanical geography, as an independent science, had already attained such development that it seemed possible and necessary to sum up the facts accumulated by it. This found expression in the publication of a number of manuals on phytogeography, in most of which problems of historical plant geography occupied a place. Among such manuals we may mention the "Handbuch der Pflanzengeographie" by DRUDE, a renowned investigator in the field of botanical geography. In this "Handbuch", which appeared in 1890, we find entire chapters devoted to such problems as areas of distribution and the flora of islands, mountain peaks, and subtropical deserts.

Very interesting and pregnant with ideas is the book of SOLMS-LAUBACH (1905), "Die leitenden Gesichtspunkte einer allgemeinen Pflanzengeographie in kurzer Darstellung". In this book the author, among a number of other problems, discusses the manner in which a species penetrates into a new habitat, changes in equilibrium in plant distribution resulting from disturbance in habitat conditions, and insular floras. Lastly, we should not fail to mention SCHRÖTER's "Genetische Pflanzengeographie" (1st ed., 1912), which gives a very excellent, though brief, exposition of the principles of historical plant geography.

Our aim in the present chapter has not been to set forth in detail the entire history of the study of our science, historical plant geography, but only to note the principal stages in this study to serve as an introduction to subsequent chapters. The problems taken up in these chapters are, in most cases, treated in their historical perspective, each chapter constituting a synthesis of numerous works and representing, so to say, a continuation of our historical review.

Historical plant geography is a science in process of formation. Before it lies a vast field for development, and the more firmly are established its underlying principles and the more data for its upbuilding are assembled by scientific investigators, the greater will be its significance as a foundation for other disciplines.

References:

ANDERSSON, G., 1897: Die Geschichte der Vegetation Schwedens (Engler's Bot. Jahrb., Vol. 22).

BEILSCHMIED, C., 1831: Pflanzengeographie nach A. HUMBOLDT's Werke über die geographische Verteilung der Gewächse (Breslau).

BORSZOW, E., 1860: Die Aralo-Caspischen Calligoneen (Mém. Acad. Sci. St. Petersb., Sér. 7, Vol. 3, No. 1).

DE CANDOLLE, ALPH., 1855: Géographie Botanique Raisonnée, Vols. I and II (Geneva).

DE CANDOLLE, AUG. PYR., 1820: Essai Élémentaire de Géographie Botanique (*in* "Dictionnaire des Sciences Naturelles", Paris, Vol. 18).

Christ, H., 1867: Über die Verbreitung der Pflanzen der europäischen Alpenkette (Neue Denkschr. Schweiz. Ges. f. Naturw., Vol. 22).

Darwin, Charles, 1859 (1st ed.): The Origin of Species (London; last (sixth) edition, 1872).

Diels, L., 1908 (1st ed.): Pflanzengeographie (Berlin; 3rd ed., 1929).

Drude, O., 1890: Handbuch der Pflanzengeographie (Stuttgart).

Engler, A., 1872: Monographie der Gattung *Saxifraga* mit besonderer Berücksichtigung der geographischen Verhältnisse (Breslau).

Engler, A., 1879, 1882: Versuch einer Entwicklungsgeschichte der Pflanzenwelt, insbesondere der Florengebiete, seit der Tertiärperiode, Vols. I and II (Leipzig).

Engler, A., 1893: Alphonse de Candolle (Ber. d. deutsch. bot. Ges., Vol. 11).

Engler, A., 1899: Die Entwicklung der Pflanzengeographie in den letzten 100 Jahren (Humboldt–Centenarschr. der Ges. f. Erdkunde, Berlin).

Engler, A., 1901: Über neuere Fortschritte der Pflanzengeographie (Engl. bot. Jahrb., Vol. 30, No. 3).

Engler, A., 1914: Pflanzengeographie. Die Kultur der Gegenwart. III, 4. Vol. 4 (Leipzig).

Forbes, E., 1846: On the connexion between the distribution of the existing fauna and flora of the British Isles and the geological changes which have affected their area, especially during the epoch of the Northern Drift (Mem. Geol. Survey of Great Britain, Vol. I).

Gams, H., 1927: Die Ergebnisse der pollenanalytischen Forschung in Bezug auf die Geschichte der Vegetation und des Klimas von Europa (Zeitschr. f. Gletscherkunde, Vol. 15).

Gray, Asa, 1846: Analogy between the flora of Japan and that of the United States (Amer. Jour. Sci., Vol. 2).

Gray, Asa, 1858–59: Observations upon the relations of the Japanese flora to that of North America (Mem. Amer. Acad. Sci., Vol. 6).

Gray, Asa, 1872: An address by Professor Asa Gray (Delivered before the American Association for the Advancement of Science, Dubuque, Iowa, August, 1872; Proc. Amer. Assn. Adv. Sci., Vol. 31).

Hooker, J. D., 1851: On the vegetation of the Galapagos Archipelago (Trans. Linnean Soc., Vol. 20).

Hooker, J. D., 1844–60: The Botany of the Antarctic Voyage, I. Flora Antarctica; II. Flora Novae Zelandiae; III. Flora Tasmaniae (London).

Hooker, J. D., 1855: Flora Indica, Vol. I (London).

Hooker, J. D., 1862: Outlines of the distribution of arctic plants (Trans. Linnean Soc., Vol. 23).

Hooker, J. D., 1866: Lecture on insular floras (Delivered before the British Association for the Advancement of Science, Nottingham, August, 1866; Gardeners' Chronicle, Jan. 1867).

Humboldt, A., 1807: Essais de géographie botanique (Paris; German ed.: Ideen zu einer Geographie der Pflanzen, Tübingen).

Kerner, A., 1869: Die Abhängigkeit der Pflanzengestalt von Klima und Boden, ein Beitrag zur Lehre von der Entstehung und Verbreitung der Arten, gestützt auf die Verwandtschaftsverhältnisse, geographische Verbreitung und Geschichte der *Cytisus*-Arten aus dem Stamme *Tubocytisus* DC. (Innsbruck).

Komarov, V. L., 1901: Flora Manshuriae, Vol. I (Acta Horti Petropolitani, Vol. 20).

Litvinov, D. I., 1891: Geobotanical Notes on the Flora of European Russia (In Russian; St. Petersburg).

Lyell, C., 1830–32: Principles of Geology, Vols. I and II (London; 9th ed., New York, 1862).

Maximovicz, C., 1859: Primitiae Florae Amurensis (Mém. Acad. Sci. St. Petersb., Vol. 9).

Meyen, F., 1836: Grundriss der Pflanzengeographie (Berlin).

Mirbel, M., 1827: Recherches sur la distribution géographique des végétaux phanérogames dans l'Ancien Monde, depuis l'équateur jusqu'au pôle arctique (Mém. Mus. d'Hist. Nat., Vol. 19).

Nathorst, A., 1892: Über den gegenwärtigen Standpunkt unserer Kenntniss von dem Vorkommen fossiler Glacialpflanzen (Bihang till K. Svenska Vetenskaps Akad. Handlingar, Vol. 17, No. 5, Appendix 3).

Roemer, M. J., 1841: Geographie und Geschichte der Pflanzen (München).

Ruprecht, F., 1866: Geobotanical Studies of the Black-Soil Zone (In Russian; Suppl. to Vol. 10, Mém. Acad. Sci. St. Petersb., No. 6).

Schouw, J. F., 1823: Grundzüge einer allgemeinen Pflanzengeographie (Berlin; Danish ed., 1822, Copenhagen).

Schröter, C., 1912: Genetische Pflanzengeographie (Handwörterbuch d. Naturwissensch., IV. Jena; 2nd ed., Jena, 1934).

Solms-Laubach, H., 1905: Die leitenden Gesichtspunkte einer allgemeinen Pflanzengeographie in kurzer Darstellung (Leipzig).

STEENSTRUP, J., 1842: Skovmoserne Vidnesdam og Lillemose (Danske Vid. Selsk. naturw. Afhandl., Vol. 9).

STROMEYER, F., 1800: Commentatio inauguralis sistens historiae vegetabilium geographicae specimen (Göttingen).

STUR, D., 1860: Beiträge zu einer Monographie des Genus *Astrantia* (Sitzungen d. math.-naturw. Klasse d. Akad. Wiss. Wien, Vol. 40).

SUKACHEV, V. N., 1937: On fossil plant remains in loess deposits in connection with their origin (Comptes Rendus (Doklady) Acad. Sci. URSS, Vol. 15, No. 4).

THISELTON-DYER, W., 1909: Geographical Distribution of Plants (*In* SEWARD: "DARWIN and Modern Science", Cambridge).

UNGER, F., 1851: Die Pflanzenwelt der Jetztzeit in ihrer historischen Bedeutung (Denkschr. d. Wiener Akad. d. Wissensch.).

UNGER, F., 1852: Versuch einer Geschichte der Pflanzenwelt (Vienna).

WAGNER, M., 1868: Die Darwin'sche Theorie und das Migrationsgesetz der Organismen (Leipzig).

WAGNER, M., 1880: Über die Entstehung der Arten durch Absonderung (Kosmos, Nos. 1, 2, 3).

WALLACE, A. R., 1876: The Geographical Distribution of Animals, Vol. I (London).

WALLACE, A. R., 1880: Island Life (London).

WILLDENOW, C., 1792: Grundriss der Kräuterkunde (Berlin; 2nd ed., Vienna, 1798).

WULFF, E. V., 1936: Significance of ALEXANDER HUMBOLDT's works for phytogeography (In Russian; *in* Russian trans. of HUMBOLDT's "Ideen zu einer Geographie der Pflanzen", Moscow-Leningrad).

Chapter III

AREAS, THEIR CENTERS AND BOUNDARIES

Area Concept Defined: — By area (*area geographica*) is understood, taking the Latin meaning of this word, the region of distribution of any taxonomic unit (species, genus, or family) of the plant (or animal) world. A distinction is made between *natural areas*, occupied by a plant as a result of its dispersal caused by the combined action of various natural factors, and *artificial areas*, arising as a consequence of the intentional or accidental introduction of a plant by man.

Within the limits of its area a plant does not occupy the entire surface of the earth but leaves smaller or larger intervening spaces unoccupied. This is due to the biological peculiarities of plants and to their adaptation to local habitat conditions, which even within the

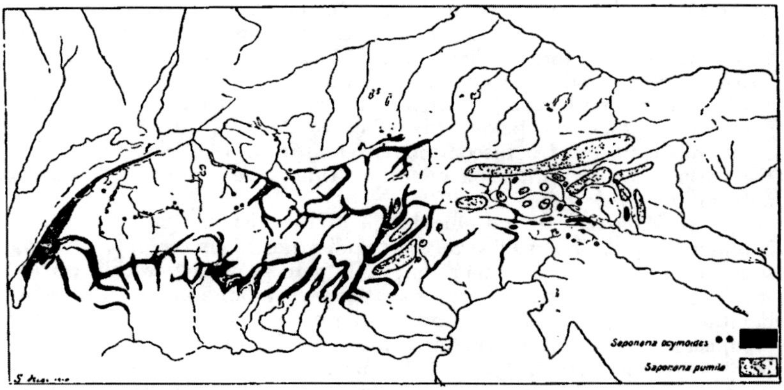

FIG. 1. — Example of a discontinuous area. Areas of *Saponaria ocymoides* and *S. pumila*, restricted to mountain-tops in the Alps. (After HEGI).

limits of a small territory may vary to a considerable extent. Among such local conditions are: physical and chemical properties of the soil and its humidity, micro-relief, micro-climate, geographical location with reference to the countries of the world, influence of animals and man, interrelations with other plants, etc. The character of the distribution of a plant within the limits of its area or, in other words, the local distribution of a plant, is known as its *topography* (DE CANDOLLE, 1855).

The area of distribution of a plant is best pictured by maps, on which all its known habitats are indicated by dots. Connecting by a line all the outer points of the distribution of a given plant, we are able to judge as to the *shape* of its area. The shape of an area depends on the combined effect of the biological peculiarities of the plant and the physico-geographical conditions of the country, the latter usually playing the predominant rôle. The configuration of an area depends, to a large extent, on the latitude. In the frigid and temperate zones, as DE CANDOLLE pointed out, the diameter of most areas from west to east is much greater than that from north to south, due to the considerably greater variation in climatic, particularly temperature, conditions, in the case of the latter direction. Such areas, therefore, have the shape of an ellipse extending from west to east. The areas of species in the torrid zone have a relatively longer (as compared to the preceding case) diameter from north to south. Cases in which both diameters are of the same length, the area being roughly circular in shape, are of very rare occurrence.

The establishment of regularities in the formation of areas and a study of the areas themselves lead to the elucidation of their history and origin, a basic task of historical plant geography, since it enables us to arrive at conclusions as to the history and origin of floras. The carrying out of this task is by no means easy. Difficulties arise from a number of sources. The chief is our insufficient knowledge regarding the flora of many regions of the globe and, hence, regarding the present geographical distribution of species, a factor which taxonomists have begun only recently to take into account in a general way. Our knowledge of the former distribution of species, since the finding of well-preserved fossil remains is of exceptionally rare and chance occurrence, is even more meager, often practically negligible. And it happens, as we shall see below, that the structure of an area in many cases, particularly of a discontinuous area, may be explained only on the basis of its conformation in former times and not on the basis of natural causes now in force.

Another obstacle to an elucidation of the distribution of a species is our inadequate knowledge of the peculiarities of its ecology. Among such peculiarities we may mention: ability to grow only within certain restricted temperature limits—*stenothermy;* adaptability to specific habitat conditions—soil, humidity, light, cohabitation with other organisms (symbiosis, mycorrhiza, parasitism), presence of special insect pollinators—*stenotopy*. Lastly, in most cases we do not know whether the given area represents the limit of possible distribution of the species or whether the area is still in the process of expansion.

As an example of the close relation between plant distribution and definite edaphic conditions we may mention plants found only on serpentine soils (LÄMMERMAYR, 1926, 1927; NOVÁK, 1928). However, as early as 1865 NAEGELI pointed out that the character of the distribution of a plant cannot be explained only by the physical or chemical nature of the soil, since the latter factor acts in combination with climatic and biotic factors. When growing apart from one another, species may be indifferent to soil conditions, while when growing together, the same species, due to mutual competition, may show preference for definite and different soils. Thus, *Rhododendron hirsutum* and *R. ferrugineum*, when found apart, grow both on soils rich and on soils poor in lime, but when found together, the former is adapted to calcareous and the latter to non-calcareous soils. Hence, edaphic conditions themselves constitute in this case only an indirect cause of the "adaptation" of these species to different soils, such "adaptation" being determined by competition between the two species.

Center of an Area: — Of vital importance for the study and understanding of an area is the determination of that initial territory whence a genus or species began its dispersal whereby it reached the present boundaries of its area. This initial territory where an area originated is known as the center of the area. One of the first definitions of the concept of the center of an area was given by ROBERT BROWN (1869),

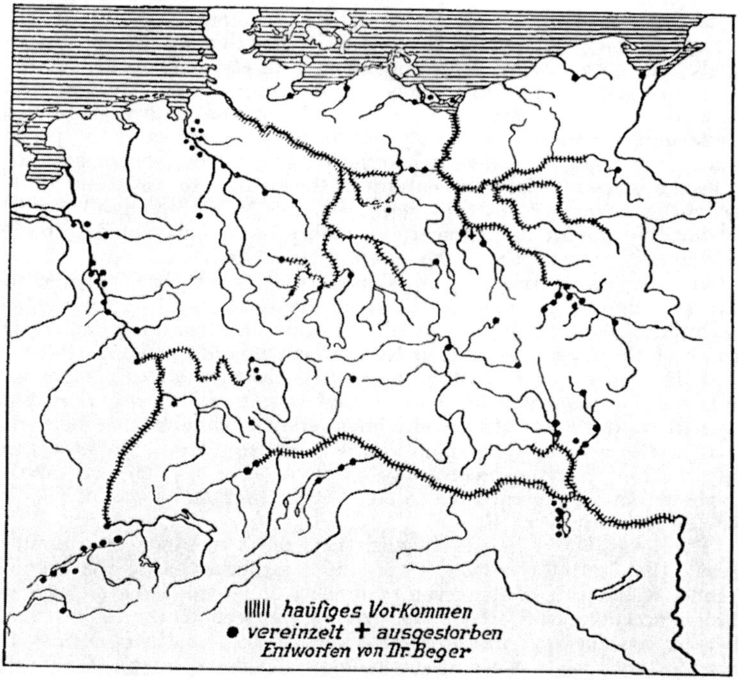

FIG. 2. — Example of a discontinuous area. Localities of *Euphorbia palustris*, restricted to the river basins of central Europe. (After HEGI).

who formulated it approximately as follows: Each genus seems to have arisen in that center in which the greater number of its species is found; these centers have doubtless undergone many modifications as a result of geological changes, and many anomalies in the distribution of plants may be thus explained.

There are no grounds for presuming that a new species will not extend its area beyond the limits of the region of its origin. It will, without any doubt, begin to spread in all directions open to it, and the region of its origin will constitute the center of the area being formed. Further on we shall discuss more in detail the origin of an area, and we shall then refer again to the problem of the initial center of an area. We shall now examine the methods of locating the center of an area.

The determination of the location of the center of an area is closely connected with the establishment of the habitats of the species or other taxonomic unit whose area is under study. Consequently, the historico-geographical study of an area should be based on a monographic study of the given taxonomic unit—species, genus, or family—and the elucidation of its kinship to closely related taxonomic units of the same or different rank.

In order to establish the center of formation of an area and the successive stages of its development, it is necessary to know its past history, which paleobotany alone is in a position to give. Unfortunately, its data are very incomplete and rarely enable one, on their basis, to establish the entire past history of an area. Nevertheless, there are very few cases when the fossil remains of any given taxonomic unit are found exclusively within the boundaries of the present area of that unit, *i.e.*, cases when we might consider the present area to be the place of origin and habitat of the given taxonomic unit during the entire period of its existence. Usually fossil remains are found outside the present area, sometimes embracing an area of distribution considerably larger than the present one and occupying entirely different regions. In such cases even very incomplete paleobotanic data give us indications as to the past area and guard us from falsely interpreting facts of the present distribution and from determining the center of the area only on the basis of contemporary data. Paleobotanic data have in many cases shown that the habitats of a taxonomic unit and, consequently, also the initial center of its area, may have been situated outside its present boundaries, a circumstance occurring both in the geography of plants and animals. Hence, in most cases only the center of the present-day distribution of a given unit may be found within its present area but not the center of its origin, *i.e.*, not the center of the area itself.

Theoretically we can distinguish between two kinds of centers of areas: the first, the region where there is accumulated the greatest number of habitats of the given taxonomic unit—the *center of frequency* (Frequenzcentrum—Samuelsson, 1910); the second, the region where there is concentrated the greatest diversity and wealth of forms—the *mass center* or *center of maximum variation* (Turrill, 1939). The latter center of an area, taking into account our present, insufficiently detailed knowledge of wild species, may be located, for the most part, only for units of the higher ranks. There is no doubt, however, that for the elucidation of the origin of an area it is of more importance to locate the center of maximum variation of the taxonomic unit whose area is under study than to locate its center of frequency, which depends more on ecological than on historical causes.

Both as regards the variety and frequency of stations and as regards the concentration of diverse forms, we may consider that a variety or species newly arising from an initial form will be found in greatest numbers not far from the place of its origin, its representatives gradually decreasing in number as one proceeds from this center of the area toward its periphery. At the time of its origin a species naturally finds itself in favorable conditions, since otherwise there would be no occasion for its arising there. If a species should accidentally arise in unsuitable conditions, it would be immediately destroyed as a result of natural selection. A new species is highly variable, reacting to all the micro- and macro-conditions of its habitat, and, hence, gives rise to a large number of forms.

PACHOSKY (1921) considers that the above-noted decrease in number of representatives of a species or variety toward the periphery of its area is closely connected with adaptability to definite habitat conditions, particularly to a definite type of soil. In the center of an area, where, as a rule, the habitat conditions of a given species most nearly approximate the optimum, it can grow under fairly diverse conditions, even on different soils. On the other hand, farther from this region, *i.e.*, nearer to the periphery of the area, not only is an optimum combination of factors of more rare occurrence but often there is lacking

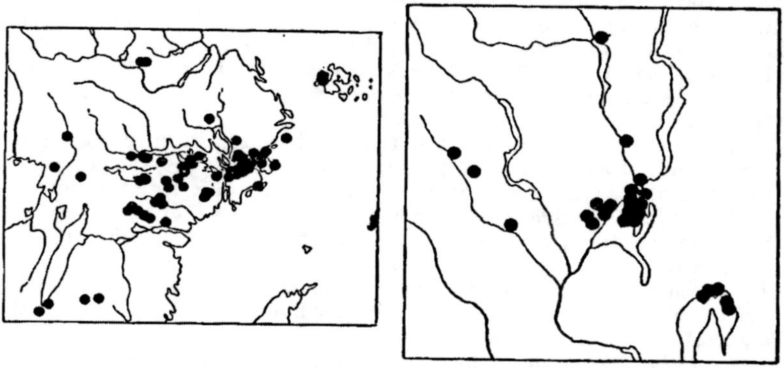

FIG. 3. — Centers of frequency in areas of species of the genus *Hieracium* on the Scandinavian peninsula (Swedish West Coast): *left, Hieracium meliceps; right, H. chloroleucum.* (After SAMUELSSON).

even that minimum of conditions required for the normal existence of a species. For example, the beech tree, ordinarily capable of growing on a variety of soils, on the periphery of its area is confined solely to lime soils. Relic species and species becoming extinct likewise prefer localities with lime or chalk soils for their habitats. Apparently, the physical conditions of the substrata of these soils provide for such species more favorable conditions as regards competition with other species, thus allowing them to maintain themselves, despite the fact that the habitat conditions as a whole deviate considerably from those normal for them.

As regards the effect of climatic factors on the distribution of a species within the limits of an area, GRAEBNER (1910) distinguishes between the "region of compact distribution", within which asp ecies finds itself in optimum conditions as regards climatic factors and the "absolute limit of distribution", where the stations of a species are confined to certain localities having specific habitat conditions.

From the foregoing it is clear that the farther from the center of an area the more rarely do conditions suitable for the growth of a given species occur, which results in the peripheral regions of an area being more sparsely inhabited by the species than the center. Moreover, plants growing under conditions unsuitable for them will quite naturally find themselves subjected to competition and crowding out by closely related species for which these same conditions are more suitable.

This, however, can by no means be taken as an unconditional and universal proposition. We can assume also the occurrence of such cases—and they actually do occur—where a species, spreading in the direction of the periphery of its area, encounters, often far from the place of its origin, favorable habitat conditions, perhaps even more favorable than existed in the center of the area, which give an impetus to new form-genesis. But such form-genesis, leading even to the origin of new species on the periphery of the area of the initial species, may also be due to unfavorable conditions. We shall discuss this in more detail later. We should, therefore, distinguish between the center of origin of an area and the center of its development, the latter in such cases being necessarily regarded as a *secondary mass center* (or centers, since there may be several of them) of the area.

Hence, when a species arises at any point of its future area there is created, first of all, a center of propagation of this species, the center of frequency of its area, and then there develops its differentiation into a number of forms of different taxonomic rank, the creation of a center of maximum variation, or a mass center of the area. In young species the latter center may not exist at all, or it may coincide with the center of frequency. Later on, such coincidence will be broken, since the primary mass center will remain in its original place, while the initial center of frequency may disintegrate and arise anew at one or another point in the migration of forms issuing from the primary center of formation of a species. A species during the course of its dispersal may, under especially favorable conditions, enter into a phase of new form-genesis, as a result of which there will arise a secondary mass center of the area, which in contrast to the primary center of its origin constitutes a center of the subsequent development of the area and will be characterized by the presence of younger forms. The secondary nature of such a center may be established by a combination of various methods of botanical study, by which it would be shown that in the center of origin there is a concentration of more ancient and primitive forms, as compared with those concentrated in the secondary center of development.

From the foregoing we may conclude that the region of frequency of stations may ordinarily be expected to coincide with the region of maximum variation, *i.e.*, with the center of origin of the area, in those cases where the distribution of the given taxonomic unit has not yet been subjected to any later influences inducing alterations in the character of the area.

As regards the area of a species, correspondence of the center of frequency with the center of the area is characteristic, as we noted above, for areas of *young species*. To illustrate this point we may present data obtained by SAMUELSSON (1910) from a study of the areas of several species endemic to the Scandinavian peninsula. His maps of the distribution of *Hieracium meticeps* and *H. chloroleucum* give a clear picture of the character of the areas of these species and the centers of greatest frequency of their stations. These centers of frequency are at the same time the centers of origin of the areas of these species. SAMUELSSON considers these species to be of comparatively recent origin, having arisen during the Ice Age after the end of the last (Mecklenburg) glaciation. Each arose—possibly by mutation—at a certain point, which became the center of its area, whence dispersal proceeded in various directions. The present boundaries of these areas, therefore, cannot be regarded as climatic boundaries. They merely mark in each case the limits of that territory which the given species has succeeded in occupying at this stage of its dispersal, a territory which in the future will continue to expand.

With respect to the terminology of the concepts of the center of an area, ARWIDSSON (1928) has proposed that those areas entirely included within the limits of one well-defined region be called *unicentric*, in contrast to *bicentric* (embracing two regions), *tricentric*, etc., or *polycentric* areas (CHRIST), if there are many such centers.

As an example of how the center of maximum variation of a genus may be located—without giving, however, any indication as to the primary or secondary nature of this center—we may take the data of SHIRJAEV (1932) on the area of distribution of the genus *Ononis*. The species and subspecies of this genus are distributed throughout the world as follows:

Country	*No. of species and subspecies*
Morocco	52
Algeria	44
Spain	44
Italy	24
Portugal	20
Syria and Palestine	19
Tunis	18
Sicily	18
Asia Minor	17
France	16
Sardinia	14
Cyprus	13
Tripolitania and Libya	13
Canary Islands	11
Mesopotamia and Kurdistan	10
Crete	10
Corsica	10
Jugoslavia	10
Islands of Aegean Sea	9
Dalmatia	9
Iran	7
Balearic Islands	7
Tyrol	7
Thrace	7
Switzerland	6
Austria	6
Istria	6
Rhodes	6
Carpathians	6
Albania	5
Egypt	5
Central Europe	4
Hungary and Rumania	4
Bulgaria	4
Arabia	4
Armenia and Transcaucasia	3
Crimea	3
Madeira	3
Greece	2
England	2
Turkestan	2
Afghanistan	2
Caucasus (excl. Transcaucasia)	1
So. European U.S.S.R.	1
Southern Siberia	1
Mongolia	1
Northwestern India	1
Eritrea	1
Abyssinia	1

From a study of these data SHIRJAEV draws the conclusion that the center of origin and development of the area of the genus *Ononis* comprised the Iberian Peninsula (Spain, 44 species; Portugal, 20 species), Morocco (52 species), and Algeria (44 species), which at one time formed a united region. This conclusion has been visualized by a map drawn by SZYMKIEWICZ (1933) on the basis of these same data.*

Another example is provided by a table given by SZYMKIEWICZ himself, where the number of species are given for regions located at approximately the same latitude from the Iberian peninsula to Japan. For each genus the figures in bold-faced type indicate the region or regions in which the genus is represented by the greatest diversity of species.

GENUS	Iberian Peninsula (so. part)	Italy	Greece	Asia Minor	Armenia & Caucasus	Iran	Soviet Central Asia	Altais	Far Eastern Region	Japan
Armeria	**37**	16	3	2	0	0	0	0	1	0
Genista	**47**	34	13	12	5	0	1	0	0	0
Helianthemum	**27**	16	10	11	7	5	1	0	0	0
Trifolium	54	**98**	64	61	45	15	14	7	2	1
Lotus	20	**24**	17	14	10	6	4	3	1	1
Coronilla	9	**11**	8	6	5	2	1	0	0	0
Silene	58	65	**86**	73	65	41	49	14	10	10
Alyssum	13	16	20	**40**	27	14	11	3	0	0
Gypsophila	3	3	7	**24**	23	16	19	7	3	0
Onobrychis	10	8	7	21	**27**	22	13	1	0	0
Astragalus	43	28	37	146	253	317	**328**	55	6	6
Ferula	5	4	3	4	9	14	**35**	3	0	0
Artemisia	20	17	5	13	20	23	**68**	30	30	17
Saussurea	0	0	0	0	1	2	**41**	23	24	19

Similarly, we may take the distribution of wild species of *Nicotiana*, a genus including, according to data of a study of the geography of this genus made by GRABOVETSKAYA (1937), a total of 76 species. They are distributed as follows: North America, 12 (of which 7 are endemic); Central America, 14 (7 endemic); South America, 43 (39 endemic); Australia, 14 (all endemic). On the basis of these data the center of the area of this genus must be regarded as South America.

This method of determining the location of the centers of origin of genera, and on the basis of the latter, the centers of development of floras, is at the present time generally accepted, but it should, nevertheless, be emphasized that this method is only relatively reliable, in many cases leading undoubtedly to incorrect conclusions. Most genera of angiosperms originated in the Cretaceous period, some probably even earlier, in the Jurassic. Having attained at the end of the Cretaceous and beginning of the Tertiary periods a very wide distribution, most of them had at that time a considerably more limited intrageneric differentiation than at present. Intensive processes of species-formation and the initiation of geographical series of species (see below) were not begun until later, in the second half of the Tertiary period. These were induced by climatic changes, particularly decrease in humidity, and geomorphological changes—mountain-forming processes (*e.g.*, the

* Discrepancies between the figures on the map and those in SHIRJAEV's table arise from the fact that for the map SZYMKIEWICZ used only the number of species, whereas SHIRJAEV included subspecies as well.

uplifting of the Alps and the Himalayas, which radically altered the climatic conditions of the regions lying north of these mountain ranges), shifting of sea basins (formation of deserts where formerly there were seas and the formation of seas where formerly there was dry land), separation from the continents of archipelagos and islands, which formerly constituted a united whole, etc.

Consequently, the present-day concentration of species only in rare instances can reflect the actual center of origin of the genus; usually it indicates the center not of the past but of the present development of the genus. In view of this, conclusions made on the basis of data on the migration of a genus from such a center of development are founded on incorrect premises and in many cases are utilized for broad generalizations, which cannot be accepted without reservations, except upon further verification.

This explains why newer methods of determining the center of origin of a genus are being sought, but it may be taken for granted that these methods will be able to give full assurance as to the reliability of the results obtained only in case they are confirmed by paleobotanic data. In the absence of the latter, conclusions drawn solely on the basis of the present distribution of species will evoke doubts as to their validity. Szymkiewicz (1934, 1936, 1937) has in recent years made intensive studies with the aim of finding new methods of locating the centers of areas and of tracing the development of floras. The methods proposed by him may be summarized as follows:

If we take as the center of origin of a genus that region where the greatest number of its species are concentrated, we do not take into account differences in the character of the areas of the various species, as a result of which we compare figures that are phytogeographically of unequal value. Szymkiewicz divides species, as regards the character of their areas, into three categories: (1) endemic and subendemic, the latter meaning species whose areas extend only slightly beyond the boundaries of their primary natural regions; (2) species whose areas embrace, in addition, a second natural region phytogeographically identical to the first; (3) widely distributed species, in whose areas the primary natural region occupies only an inconsiderable part. These three categories of species, in judging as to the center and origin of the area of a genus, provide data of unequal value, the first being of greater significance than the second and the second greater than the third. Szymkiewicz (1937) proposes, therefore, that the center of the area of a genus should be established not on the basis merely of data as to the total number of species but of data as to the number of species in each of the three above-mentioned categories, and he points out that by the latter method it is easier to detect a second center of concentration of species, in case there are two such centers. By way of illustration are given below the data obtained by Szymkiewicz for the genus *Carex:*

Carex

Regions of distribution	No. of species in each category	Total no. of species
Europe	27–101–14	142
Siberia	8– 70–16	94
Mediterranean Basin	31– 67–22	120
Eastern Asia	259– 70–18	347
No. America, Pacific Coast	61– 77– 5	143
– – , Atlantic –	86– 75– 6	167
Mexico	9– 11– 3	23
Andes	41– 16– 3	60
Neotropical region	12– 5– 2	19
Tropical Africa	28– 6– 0	34
Malaysia	51– 17– 8	76
South Africa	5– 9– 0	14
Australia	40– 11– 1	52

This table shows that, taking the greatest concentration of species as a basis, Eastern Asia would appear to be the chief center of origin of the genus *Carex* and the Atlantic Coast of North America the next most important center. Judging by the concentration of each of the three categories of species, the conclusion that these are the two most important centers is confirmed, for they contain the greatest number of endemic species.

But in thus locating the center of an area it is necessary also to take into account the fact, well known to every author of a botanical monograph, that species themselves are not uniform, not of equal value. This may usually be compensated for by the grouping together of closely related species into sections, subgenera, and other such units. Consequently, for the purpose of checking the conclusions as to the center of the area of a genus arrived at on the basis of a calculation of the number of species in the three categories, Szymkiewicz proposes that analogous calculations be made of the number of sections or subgenera in each of the same three categories. Thus, one first calculates the number of species of each of the three categories in each section, and then, on this basis, determines the number of endemic sections, sections distributed in two regions, and sections having a wide distribution. As an example let us take the data obtained for the genus *Sisymbrium:* —

Sisymbrium

Regions of distribution	No. of sections in each category	Total no. of sections	No. of species in each category	Total no. of species
Europe	0–1–4	5	5–5–0	10
Siberia	0–0–2	2	0–3–0	3
Mediterranean Basin	5–2–1	8	15–6–1	22
Eastern Asia	0–0–2	2	1–1–0	2
No. America, Pacific Coast	0–0–2	2	2–0–0	2
Mexico	0–0–2	2	3–0–0	3
Andes	5–1–0	6	25–0–0	25
Neotropical region	0–0–1	1	1–0–0	1
Tropical Africa	0–1–1	2	2–0–0	2
South Africa	2–0–0	2	10–0–0	10

In this case by all three methods we get the same result: the existence of two centers, one in the Mediterranean Basin and one in the Andes. This connection between the Mediterranean Basin and Central America and adjoining territories of North and South America is characteristic of a number of genera, such as *Draba*, *Eryngium*, *Centaurea*, *Astragalus*, *Trifolium*, *Lupinus*, etc. But such an exact agreement of results from all three methods of investigation does not, by any means, always hold true. As an illustration of this, we may take the distribution of sections in the genus *Carex*, for each of its subgenera separately, and compare the data thus obtained by Szymkiewicz with those given above for species.

Carex

Regions of distribution	Subgenus *Primocarex*	Subgenus *Vignea*	Subgenus *Indo-Carex*	Subgenus *Eu-Carex*
	No. of sections in each category	No. of sections in each category	No. of sections in each category	No. of sections in each category
Europe	0–3–4	2–2–11	0–0–0	2–1–18
Siberia	0–1–5	0–0–11	0–0–0	0–1–15
Mediterranean Basin	0–1–4	1–1–11	0–1–0	0–0–20
Eastern Asia	2–1–0	2–2–10	2–0–3	8–4–10
North America — Pacific Coast	3–3–3	2–4– 8	0–0–0	1–3–16
North America — Atlantic Coast	2–3–4	2–4– 6	0–0–0	7–5–10
Mexico	0–0–0	0–0– 3	0–1–1	0–1– 8
Andes	1–0–2	2–0– 6	0–0–2	1–1–11
Neotropical region	1–0–0	1–0– 1	0–0–2	0–0– 4
Tropical Africa	1–0–0	0–1– 0	0–0–1	1–0– 4
Malaysia	0–0–2	0–0– 5	3–0–0	0–0–10
South Africa	0–0–0	0–0– 2	0–0–1	0–0– 6
Australia	0–0–1	2–1– 4	0–0–0	1–1– 7

In this table there does not stand out any definite center of the area of the genus, the East-Asiatic center so prominent in the other table for the genus *Carex* not being in evidence here at all. We thus see how complicated is the problem of determining the location of the center of an area, and that for its solution statistical calculations alone can have only a very limited significance. It is necessary to assemble data of an all-sided study of species—as, for instance, was done by KOMAROV (1908)—disclosing the phylogenetically most primitive types, the direction of their evolution, the centers of concentration of these primitive, initial types, and the direction of their further distribution. Only such a monographic study, based also on paleobotanic data, can give a more or less correct idea as to the initial center of the area of a genus and of the secondary centers of its development.

In this respect cytological data may prove of great value. It has now been established that in some cases species belonging to the same genus differ in chromosome number, and that many of the polyploid species originated, apparently, as a result of chromosome mutations induced by the action of external factors. Species, in dispersing from the center of their origin, often extend their area beyond the boundaries of optimum conditions for their existence. As a result of the action of ecological factors to which a species is not accustomed there occur irregularities at meiosis in the sex cells, which result in the phenomenon of polyploidy. This connection between the origin of polyploid species and definite ecological conditions is the reason why such species have in many cases quite specific geographical areas, differing from the areas of the initial species (see below; also, in more detail, WULFF, 1937).

It is now possible, therefore, theoretically to advance the proposition that floras of those regions of the globe characterized by extremely low temperature, such as arctic regions and mountain peaks, or by very high temperature and low humidity, such as deserts, are distinguished by an exceptionally large number of polyploid species. It must also be presumed that such chromosome mutations have occurred in nature not only under present-day conditions, as a result of species having become widely distributed and having penetrated into localities with ecological conditions differing from those normal for them, but also as a result of the great climatic changes that took place in former geological times and of the migrations of species in those times.

Arranging the species of a genus in order according to chromosome number, we obtain so-called polyploid series of species, which at the same time reflect the direction of evolution of the genus and also the direction of its dispersal. Starting from the premise that the species having the smallest chromosome number in a polyploid series usually is the initial species, we may consider that the areas of species characterized by such chromosome numbers are more ancient than areas of species with larger chromosome numbers and that, consequently, in the regions occupied by these ancient areas one must seek for the initial center of the area of a genus. As an illustration we may cite the data of a cytological investigation of the genus *Iris* carried out by SIMONET (1932). This genus is widely distributed throughout the entire northern hemisphere. Species having rhizomes occupy the largest areas, practically identical to that of the genus, while tuber-bearing species are considerably more restricted in their distribution. The area of the latter is confined to the Mediterranean Basin—from the Iberian Peninsula to Soviet Central Asia, inclusive. Not only all four sections of tuber-bearing irises but also three of the sections of rhizome-bearing irises are limited in their distribution to the Mediterranean Basin. Moreover, almost all the other sections have representatives here. These circumstances force one to presume that the center of origin of the area of the genus *Iris* must be located in the Mediterranean Basin. This conclusion is confirmed by cytological data. Precisely in the Mediterranean Basin are concentrated those species with the lowest chromosome numbers (n = 8, 9, 10, 11), whereas American species with the highest chromosome numbers have areas located at the greatest distance from this center. Moreover, the tuber-bearing species of irises, being the most ancient species and having an area confined to the Mediterranean Basin, have the lowest chromosome numbers. Hence, if, in determining the location of the centers of areas, we utilize cytological data for those genera the species composing which have different chromosome numbers, we acquire an additional method facilitating, in combination with other methods, the solution of this difficult problem.

If, after a species has died out over a considerable portion of its area, favorable conditions should reoccur, the species may renew its dispersal from those retreats where it preserved its habitats. For example, many species lived through the Ice Age in restricted localities, which served as retreats for them and whence, in inter- and post-glacial periods, they renewed their dispersal. In such cases these retreats are known as *centers of dispersal* (centres de dispersion—Jeannel and Joleaud, 1924) or *centers* (*regions*) *of preservation* (Erhaltungsgebiete—Irmscher, 1929).

In determining the location of the center of an area great caution must be observed, since if any factors whatsoever favoring or hindering the distribution of species are not taken into account, entirely incorrect conclusions may be drawn. For instance, Palmgren (1927), on the basis of the character of the distribution of species on the Åland Islands, draws conclusions as to the extent of their penetration into the territory of these islands and the general direction of their migration. He considers that, in case a species is distributed *uniformly* within the limits of a given territory, we cannot obtain any facts as regards its former migrations from its present distribution. By uniform distribution he means the approximately equal frequency of occurrence of a species in *all* parts of the given territory and, consequently, the absence of any perceptible concentration of stations in any one part. If, on the other hand, the frequency of occurrence of a species grows clearly less or greater in some definite direction, this indicates, in his opinion, the direction of migration of the species. Thus, by his investigation of the Åland Islands Palmgren found that, in addition to uniformly distributed species, there are three other categories of species, which, in contradistinction to the former, shed some light on their origin on the islands. The first category embraces species with a clear decrease in the frequency of their occurrence toward the east, which gives grounds for considering that they migrated from the west. This group of species is the largest in point of numbers in the flora of the Åland Islands. To the second category belong a few species, the frequency of occurrence of which decreases toward the west, and which, consequently, migrated presumably from the east. Lastly, the third category embraces species distributed within the limits of the islands in two isolated areas—western and eastern. From these data PALMGREN draws the conclusion that the first category of species, constituting the great bulk of the flora of the islands, migrated to the islands from the Scandinavian peninsula, from Sweden; the second category from Finland or the eastern section of the Baltic seacoast; and, lastly, the third category from both directions.

But, in opposition to the foregoing, EKLUND (1931) shows that in southwestern Finland there is found a very great diversity of habitat conditions, particularly of edaphic conditions. This diversity is very clearly reflected in the distribution of plants. In the western part of this region there are the best soil conditions, shown, first of all, in the fairly high content of lime in the soil. Here the flora is richest. From this locality in all directions the soil grows poorer, accompanied by an impoverishment of the flora. Hence, the decrease in the frequency of occurrence of species from west to east is to be explained not by the greater distance from the place from which they migrated, as PALMGREN assumed, but by ecological causes, expressed in this case by the indicated differences in soil conditions. EKLUND remarks that in the Åland Islands there may be observed a decrease in the frequency of occurrence of species and an impoverishment of the flora from west to east, while in Uppland there is just the reverse—an impoverishment from east to west. In both cases this impoverishment is to be explained by one and the same cause, by a decrease in the content of lime in the soil. Consequently, PALMGREN's conclusions with respect to the direction of migration and the chief country from which the Åland Islands derived their flora, based on the decrease in frequency of occurrence of species from west to east, are in the given case incorrect, since he did not take ecological conditions into account. EKLUND comes to the conclusion that the islands were populated with species from an entirely different direction than PALMGREN supposed.

The present areas of many species do not constitute the maximum territory that they may possibly be capable of occupying. The further expansion of these areas has been curtailed by obstacles that have up to the present prevented these species from continuing to spread. By the artificial introduction of plants into new habitats outside their natural area it is frequently found that a species may grow under a considerably wider range of ecological and geographical conditions. This shows that each species has, besides its actual area, a "potential area" (GOOD, 1931). This circumstance is of exceptional practical importance in the introduction and regional allocation of new crops.

Boundaries of an area: — The limits of distribution of a species, the *boundaries of its area*, formerly very ineptly termed "vegetation lines", are determined by the reaction of the species to any of numerous factors or combinations of these factors. Among the most obvious causes hindering the dispersal of a species are purely mechanical obstacles, such as mountains, seas, deserts, etc. Only in rare cases does a plant, by the mere dissemination of its seeds and their transport by chance agents, succeed in overcoming such obstacles and extending its area beyond them.

Among other factors limiting the extent of an area—and for the plant usually insurmountable—are climatic conditions. The latter, creating the *climatic boundaries* of an area, may limit the distribution of a species both horizontally (to the north, south, east, and west) and vertically (altitudinally). Climatic boundaries are not determined by any one climatic factor but by all of them taken together, in consequence of which a study of the climatic boundaries of an area and the elucidation of the rôle of individual climatic factors meet with very great difficulties. The latter are all the greater because the reaction of species to climatic phenomena is closely linked with their biological characteristics, as a result of which their climatic boundaries are characterized by extraordinary diversity. Nevertheless, a study of the areas of plants provides a basis for determining the most important climatic factors affecting their distribution.

Altitudinal climatic boundaries are the result of a particularly complex combination of causes, often very difficult to fathom, the most important of all being insufficient warmth (inadequate sum of temperatures above the minimum temperature required for the given species). In addition to the latter, insufficient humidity, intensity of the sun's rays at high altitudes, strong heat radiation, eternal snow or late melting of snow, height above sea level depending on the latitude of the locality, and other factors also play a part in determining the altitudinal climatic boundaries of an area.

Until the end of the nineteenth century evaluations of climate and also of the climatic boundaries of the areas of plants were made chiefly on the basis of temperature data. At the beginning of the twentieth century it became clear that atmospheric humidity, as a factor determining the boundaries of the distribution of plants, was of predominant importance. In many cases, as, for example, for most evergreen plants, temperature plays only an indirect rôle in the limitation of the distribution of a species, the chief factor being humidity conditions. Hence, in determining the boundaries of areas, both these factors should be taken into account (GAMS, 1931).

The boundaries of an area may be determined not only by climatic causes but also by edaphic causes or by a combination of edaphic, climatic, and geographical causes. Lastly, competition with other plants may create an insurmountable obstacle to the further distribution of a species.

In many cases the boundaries of areas cannot be explained by any cause at present in force, due to the fact that these areas were formed under the influence of conditions in past epochs, often in other geological periods. A study of such areas and the establishment, on the basis of such a study, of the history of distribution of a given species constitute one of the chief tasks of historical plant geography.

The boundaries of areas may, then, be subdivided into three main types: first, boundaries set by physical barriers impassable for the given species, such as seas, straits, rivers, mountains, deserts, etc.; second, boundaries determined by ecological conditions; and, third, boundaries determined by competition among species. Moreover, an area may be in a state of expansion, in case the dispersal of the species is still in progress, or, on the other hand, it may be in a state of contraction, in case of retrogressive distribution. In the latter case the contraction of an area may for a time be in abeyance and the boundaries of the area remain temporarily without change, but subsequently they may either continue to contract or, in case of the onset of more favorable conditions, begin again to expand.

Areas vary greatly in size, depending on a combination of factors, among which the history of the given species plays an important rôle. If we assume that an area has a center of origin, from which there took place the gradual dispersal of a species or other taxonomic unit in different directions, it seems necessary likewise to assume that the size of the area occupied, in case of unhindered dispersal, would depend in part on the duration of such dispersal, which may be designated as the "age" of the species. Thus, SCHULZ (1894) considers that only very few species of the flora of central Europe have succeeded in attaining in post-glacial times, and these only in a few places, their natural boundaries as set by their edaphic and climatic requirements and by their ability to spread.

In botanico-geographical literature age as a factor in plant distribution has long been recognized. As early as 1853 LYELL in his "Principles of Geology", in chapters on the distribution of plants and animals, wrote that, if we assume that a species arises only in one place, it must have considerable time to become distributed over an extensive area. If this hypothesis is accepted, it follows that restricted distribution may, in the case of some species, be due to their recent origin and, in the case of others, to the fact that the area they once occupied has been greatly contracted as a result of climatic changes. The former are young, local species that have not existed long enough to have had the possibility for widespread dispersal, while the latter are no doubt of considerable age.

HOOKER, in his "Flora Novae Zelandiae" (1853), writes that "consistently with the theory of the antiquity of the alpine flora of New Zealand, we should find amongst the plants common to New Zealand and the Antarctic Islands some of the most cosmopolitan, and we do so". But, at the same time, HOOKER, fully conceding that all the diversity in the geographical distribution of plants cannot be explained by age alone, goes on to say that ". . . though we may safely pronounce most species of ubiquitous plants to have outlived many geological changes, we may not reverse the position, and assume local species to be among the most recently created, for species, like individuals, die out in the course of time; whether following some inscrutable law whose operations we have not yet traced, or whether . . . they are destroyed by natural causes (geological or other) they must in either case become scarce and local while they are in process of disappearance" (p. xxv).

An equally clear exposition of the significance of age as a botanico-geographical factor may be found in BENTHAM's "Notes on the Classification, History, and Geographical Distribution of the Compositae" (1873).

SCHRÖTER (1913) points out that the degree of disruption may also be utilized in determining the age of an area, an extensive and much disrupted area indicating its considerable age. POHLE (1925) proposes that ancient species be called "senior species" and young species—"junior species".

These citations fully suffice to show that the significance of age was never lost sight of by botanical geographers, although, no doubt, as compared with other biological factors, it was given too little attention and its importance underrated.

Age—or the length of time during which the dispersal of species and, hence, also the formation of floras have taken place—is for historical plant geography a factor of just such prime importance as the duration of geological periods established by LYELL was for the theory of evolution. Just as a necessary premise for the evolution of organisms is the duration of time taken by the latter for their development, so all the regularities in historical biogeography may be understood only by taking into account the length of time during which they have existed.

Consequently, we cannot fail to give serious consideration to WILLIS's treatise on "Age and Area" (1922), devoted to a study of it as a botanico-geographical factor and constituting a summary of his many investigations on this problem begun in 1907. The works of WILLIS are not mere armchair theorizing but are based on twenty years of field work devoted to the geographical study of plants in nature, in tropical South America and particularly in tropical Asia, where for a long time he was director of the Botanical Garden at Ceylon. He verified his conclusions by comparison with the data published by him in his "Flora of Ceylon" and other floristic works. At the same time, however, the very fact that he used the floras of these tropical regions as the basis for his conclusions constitutes the cause of the onesidedness of his conclusions, on account of which numerous criticisms were directed against him. The floras of the tropical regions of America and Asia are the only floras on the globe that since the Cretaceous period have not been subjected to great climatic changes. Consequently, laws established with respect to the formation of areas of species of these floras are applicable only to such floras as are characterized by unhampered development. They are not of universal significance and cannot be applied to all the floras of the earth. The criticisms of WILLIS's book were directed chiefly on this flaw in the propositions advanced by him.

Studying the flora of Ceylon, WILLIS was struck by the great differences in the size of areas occupied by different species of the same genus, some of which were endemic to the island and others not. This led him to the conclusion that "the endemic species occupied, on the *average*, the smallest areas in the island, those found also in Peninsular India (but not beyond) areas rather larger, and those that ranged beyond the peninsula the largest areas of all (again on the average)" (WILLIS, 1922, p. 65). At the same time, the number of species in each class was found to vary, increasing or diminishing depending on the size of the area. This may be clearly illustrated by his data on the flora of New Zealand. Taking the extent of the areas of species in this flora along the north and south diameter of the island, the following gradations are obtained (*ibid*., p. 64):—

	Range in N. Z. (miles)	Endemics	Wides
1.	881–1080	112	201
2.	641– 880	120	77
3.	401– 640	184	53
4.	161– 400	190	38
5.	1– 160	296	30*

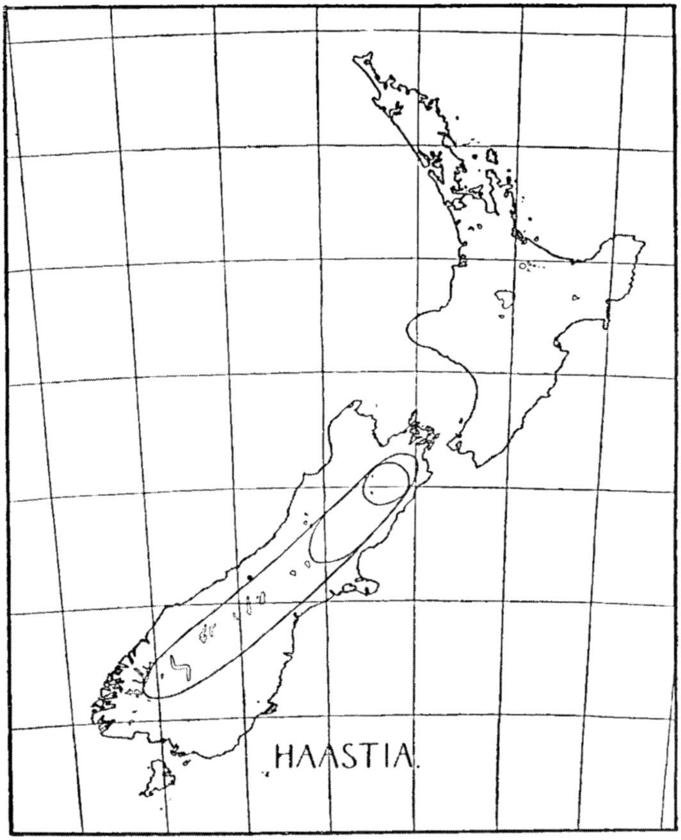

Fig. 4. — Areas of species of different ages, the most ancient species occupy the largest areas: species of the genus *Haastia* (*Compositae*) in New Zealand. (After Willis).

The widely held view that endemic species are either relics approaching extinction or species that have arisen as a result of adaptation to local conditions cannot explain the fact of gradual gradation in the areas occupied both by endemics and by wides, the first from many

* Largely undoubted introductions of recent years.

small areas to few large and the second in the reverse order. To explain this regularity it is necessary to concede the significance of age as a factor in distribution. The older species with extensive areas of distribution reached New Zealand prior to its separation from Australia and had enough time to become widely distributed there. Hence, it is clear why in the zone of areas of least extent we find the smallest

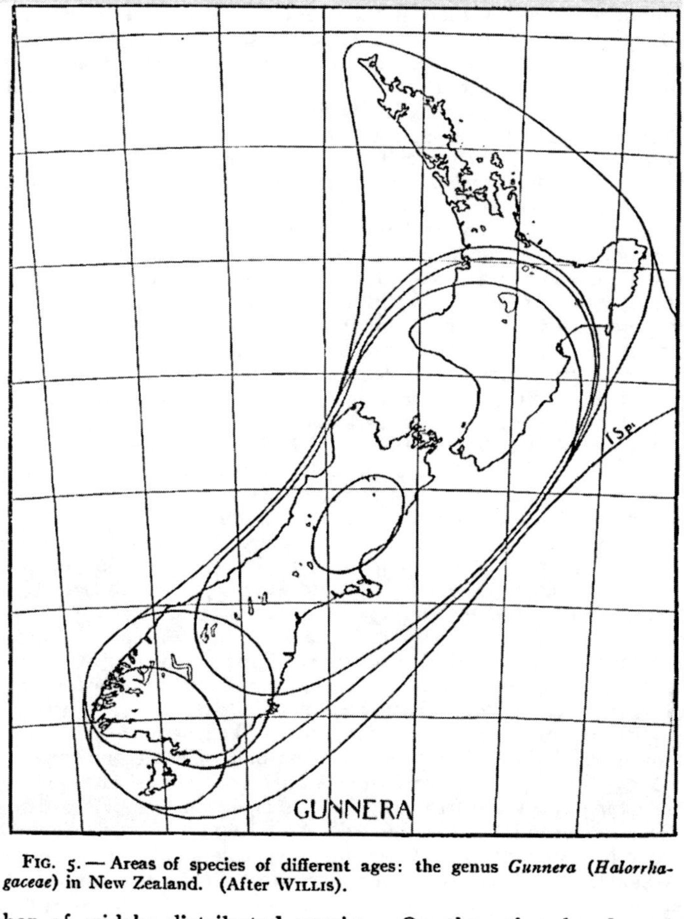

FIG. 5. — Areas of species of different ages: the genus *Gunnera* (*Halorrhagaceae*) in New Zealand. (After WILLIS).

number of widely distributed species. On the other hand, endemic species—regarded by WILLIS as young species that had their origin at a later time, after New Zealand had become an island—become more and more rare, the greater the distance from the place of their origin. Consequently, it was to be expected, and investigations confirmed this, that the islands surrounding New Zealand would have a flora consisting of the oldest species, those most widely distributed in New Zealand. To quote WILLIS (1922, p. 75): "In fact it was found that on the average its species ranged nearly 300 miles more in New Zealand than did those that did not reach the islands".

As regards the way in which a country is peopled by invasions of plants, WILLIS gives the following rule based on his general hypothesis (*ibid.*, p. 83): "If a species enter the country and give rise casually to new (endemic) species, then, if the country be divided into equal zones, it will generally occur that the endemic species occupy the zones in numbers increasing from the outer margins to some point near the centre at which the parent entered". Applying this to New Zealand, he found that all the genera in its flora adhered to this rule.

Supplementary to his central "Age and Area" hypothesis, WILLIS proposes a second principle called by him "Size and Space", which he formulates as follows (*ibid.*, p. 118): "If species spread in a country mainly in accordance with their age, then it is clear that on the average some of those in the genera represented by most species will have arrived before the first of those in the genera represented by few; . . . on the whole, keeping to the same circle of affinity, a group of large genera will occupy more space than a group of small. The space occupied will vary more or less with the number of species".

From the foregoing follows also the final implication of WILLIS's theory, *viz.*, that monotypic genera, that is, genera with one species only, like endemic genera and species, are "young beginners" that have just commenced their geographical distribution. Here, however, one must make the reservation (which WILLIS himself fails to make) that this conclusion is not applicable to genera that have acquired their monotypic character as a result of the dying out of most of their species nor to ancient endemic genera and species.

All of WILLIS's views may be summarized in this basic hypothesis—the area of a species is proportional to its age. If this proposition could be universally applied, it would simplify the solution of many problems of botanical geography. But, as his critics pointed out, this proposition is applicable only to certain genera and species, and so it cannot serve as a general rule for determining the age of an area. For instance, paleobotanic data show that some genera now occupying small areas were widely distributed in the past and are often older than genera now having extensive areas. The same applies to those endemic species that are the descendants of species at one time widely distributed but whose areas were much contracted (BERRY, 1917). In his later papers and in his book WILLIS, in answer to the deluge of criticisms, reformulated his Age and Area hypothesis, qualifying it by so many reservations that it became very complicated and practically unworkable. In its latest version it read: "The area occupied at any given time, in any given country, by any group of allied species at least ten in number, depends chiefly, so long as conditions remain reasonably constant, upon the ages of the species of that group in that country, but may be enormously modified by the presence of barriers such as seas, rivers, mountains, changes of climate from one region to the next, or other ecological boundaries, and the like, also by the action of man, and by other causes" (WILLIS, 1922, p. 63).

The areas of species of a number of floras (of England—GUPPY. 1925; MATTHEWS, 1922; of North America—FERNALD, 1924; of South Africa—SCHONLAND, 1924; and others) have been studied with the aim of testing WILLIS's theories, and the investigators came to the conclusion that these theories were not applicable to the cases studied by them. The size of the areas studied depended not so much on their age as on the adaptability of the given species and on whether or not ecological conditions favored dispersal. Species that had migrated at a later period often had larger areas than older species, parts of the areas of which had been destroyed during the Ice Age. The region of greatest concentration of endemics did not coincide with the place of origin of the genera within the limits of the given flora. Moreover, it was shown that WILLIS entirely ignored those changes in the composition of floras induced by man's activities (RIDLEY, 1923). Likewise studies of various families and genera (*e.g.*, *Magnoliaceae*—GOOD, 1925; *Passerina*—THODAY, 1925) also revealed a number of data disagreeing with WILLIS's theory. On the other hand, several investigators have presented data that agree with the regularities established by him.

All this indicates that the size of an area does not depend solely on the age of a species. The latter constitutes only one of a combination of factors on which area-formation depends. Nevertheless, the study of age as a factor in plant distribution, the significance of which was first emphasized by WILLIS, should be continued.

References:

ARBER, AGNES, 1919: On the law of age and area in relation to the extinction of species (Ann. Bot., Vol. 33, pp. 211–213).

ARWIDSSON, TH., 1928: Bizentrische Arten in Skandinavien—eine terminologische Erörterung (Bot. Notiser, No. 1).

BENTHAM, G., 1873: Note on the classification, history and geographical distribution of the *Compositae* (Linn. Soc. Jour., Vol. 13).

BERRY, E. W., 1917: A note on the age and area hypothesis (Science, Vol. 46, pp. 539–40).

BERRY, E. W., 1924: Age and area as viewed by the paleontologist (Amer. Jour. Bot., Vol. 11, No. 9).

BROCKMAN-JEROSCH, H., 1913: Der Einfluss des "Klimacharakters" auf die Grenzen der Pflanzenareale (Vierteljahrsschr. Naturf. Ges. Zürich, Vol. 58).

BROCKMAN-JEROSCH, H., 1913: Der Einfluss des "Klimacharakters" auf die Verbreitung der Pflanzengesellschaften (Engler's Bot. Jahrb., Vol. 49, Suppl. 109).

BROWN, R., 1869: On the geographical distribution of the *Coniferae* and *Gnetaceae* (Trans. Bot. Soc., Vol. 10).

BUNGE, S., 1874: Weite und enge Verbreitungsbezirke einiger Pflanzen (Sitzungsber. Dorp. Naturf. Ges. Jurjev, Vol. 11).

CHRIST, H., 1913: Über das Vorkommen des Buchsbaumes (*Buxus sempervirens*) in der Schweiz und weiterhin durch Europa und Vorderasien (Verhandl. Naturf. Ges. Basel, Vol. 24).

EKLUND, O., 1931: Über die Ursachen der regionalen Verteilung der Schärenflora Südwest-Finnlands (Acta Bot. Fenn., Vol. 8).

EKLUND, O., 1937: Klimabedingte Artenareale (Acta Soc. pro Fauna et Flora Fennica, Vol. 60).

FERNALD, M., 1924: Isolation and endemism in northeastern America and their relation to the age and area hypothesis (Amer. Jour. Bot., Vol. 11, No. 9).

FERNALD, M., 1926: The antiquity and dispersal of vascular plants (Quart. Rev. Biol., Vol. 1).

GAMS, H., 1931: Die klimatische Begrenzung der Pflanzenareale (Zeitschr. d. Ges. f. Erdkunde, No. 9/10).

GLEASON, H. A., 1924: Age and area from the viewpoint of phytogeography (Amer. Jour. Bot., Vol. 11).

GOOD, R. D'O., 1925: The past and present distribution of the *Magnoliae* (Ann. Bot., Vol. 39, No. 154).

GOOD, R. D'O., 1931: A theory of plant geography (New Phytol., Vol. 30, No. 3).
GRABOVETSKAYA, A. N., 1937: A contribution to our knowledge of the genus *Nicotiana* L. (In Russian, Eng. summary; Bull. Appl. Bot., Gen. and Plantbr., Ser. 1, No. 2).
GREENMAN, J. M., 1925: The age and area hypothesis with special reference to the flora of Tropical America (Amer. Jour. Bot., Vol. 12, No. 3).
GRIGGS, R. F., 1914: Observations on the behavior of some species on the edges of their ranges (Bull. Torrey Bot. Club, Vol. 41).
GUPPY, H. B., 1910: Die Verbreitung der Pflanzen und Tiere (Peterm. Mitt., Vol. 56).
GUPPY, H. B., 1921: The testimony of the endemic species of the Canary Islands in favour of the age and area theory of Dr. WILLIS (Ann. Bot., Vol. 35).
GUPPY, H. B., 1925: A side issue of the Age and Area hypothesis (Ann. Bot., Vol. 39, No. 156).
HULTÉN, E., 1937: Outline of the history of arctic and boreal biota (Stockholm).
IRMSCHER, E., 1929: Pflanzenverbreitung und Entwicklung der Kontinente, Parts I and II (Mitt. aus d. Inst. f. allg. Bot., Hamburg, Vol. 8).
JEANNEL, R. et JOLEAUD, L., 1924: Centres de dispersion (C. R. Soc. Biogéogr., No. 2).
KRAŠAN, FR., 1880: Über gewisse extreme Erscheinungen aus der geographischen Verbreitung der Pflanzen (Zeitschr. Oest. Ges. f. Meteor., Vol. 15).
KRAŠAN, FR., 1882: Die Erdwärme als Pflanzengeographischer Factor (Engler's Bot. Jahrb., Vol. 11).
LÄMMERMAYR, L., 1926: Materialien zur Systematik und Oekologie der Serpentinflora (Sitz. Ber. Ak. Wiss. Wien, Abt. I, Vol. 135, No. 9).
LANGLET, O., 1935: Über den Zusammenhang zwischen Temperatur und Verbreitungsgrenzen von Pflanzen (Meddel. Stat. Skogsförsöksanst., Stockholm, Vol. 4, No. 28).
MATTHEWS, J. R., 1922: The distribution of plants in Perthshire in relation to Age and Area (Ann. Bot., Vol. 36, No. 143).
NAEGELI, K., 1865: Über die Bedingungen des Vorkommens von Arten und Varietäten innerhalb ihres Verbreitungsbezirkes (Sitzungsber. d. math.-phys. Klasse d. Bayr. Akad., Vol. 2, No. 4).
NOVÁK, F., 1928: Quelques remarques relatives au problème de la végétation sur les terrains serpentiques (Preslia, Vol. 6).
PALMGREN, A., 1927: Die Einwanderungswege der Flora nach den Ålandsinseln (Acta Bot. Fennica, Vol. 2).
POHLE, R., 1925: Drabae asiaticae (Repert. sp. nov. Beih., Vol. 32).
RIDLEY, H. N., 1923: The distribution of plants (Ann. Bot., Vol. 37, No. 1).
RIDLEY, H. N., 1925: Endemic plants (Jour. Bot., Vol. 63).
SALISBURY, E. J., 1926: The geographical distribution of plants in relation to climatic factors (Geogr. Jour.).
SAMUELSSON, G., 1910: Über die Verbreitung einiger endemischen Pflanzen (Arkiv f. Bot., Vol. 9, No. 12).
SCHONLAND, S., 1924: On the theory of Age and Area (Ann. Bot., Vol. 38, No. 151).
SCHULZ, A., 1894: Grundzüge einer Entwicklungsgeschichte der Pflanzenwelt Mitteleuropas seit dem Ausgange der Tertiärzeit (Jena).
SHIRJAEV, G., 1932: Generis *Ononis* revisio critica (Beih. z. Bot. Centralbl., Vol. 49, Abt. 2).
SIMONET, M., 1932: Nouvelles recherches cytologiques et génétiques chez les *Iris* (Ann. Sci. Nat. Bot., sér. 10).
SINNOTT, E., 1917: The Age and Area hypothesis and the problem of endemism (Ann. Bot., Vol. 31).
SINNOTT, E., 1924: Age and Area and the history of species (Amer. Jour. Bot., Vol. 11, No. 9).
SZYMKIEWICZ, D., 1933: Contributions à la géographie des plantes, I–III (Kosmos, Vol. 53).
SZYMKIEWICZ, D., 1934: Une contribution statistique à la géographie floristique (Acta Soc. Bot. Polon., Vol. 11, No. 3).
SZYMKIEWICZ, D., 1936: Seconde contribution statistique à la géographie floristique (Acta Soc. Bot. Polon., Vol. 13, No. 4).
SZYMKIEWICZ, D., 1937: Contributions à la géographie des plantes, IV. Une nouvelle méthode pour la recherche des centres de distribution géographique des genres (Kosmos, Vol. 61).
SZYMKIEWICZ, D., 1939: Une nouvelle méthode . . . (Chron. Bot. 5:201).
THODAY, D., 1925: The geographical distribution and ecology of *Passerina* (Ann. Bot., Vol. 30, No. 153).
TURRILL, W. B., 1939: The principles of plant geography (Bull. Misc. Information, Kew, No. 5, pp. 208–237).
WILLIS, J. C., 1917: Further evidence for age and area (Ann. Bot., Vol. 31).
WILLIS, J. C., 1922: Age and Area. A study in geographical distribution and origin of species (Cambridge).
WILLIS, J. C., 1923: Age and Area: a reply to criticism, with further evidence (Ann. Bot., Vol. 37, No. 146).

WILLIS, J. C., 1923: The origin of species by large, rather than by gradual change and by GUPPY's method of differentiation (Ann. Bot., Vol. 37, No. 148).

WILLIS, J. C., 1936: Some further studies in endemism (Proc. Linn. Soc. London, Session 148, Part 2).

WULFF, E. V., 1936: Area y Edad (Rev. Argent. de Agronom., Vol. 3, No. 1; Pub. in 1927 in Russian in Bull. Appl. Bot., Vol. 17, No. 4).

WULFF, E. V., 1937: Polyploidy and the geographical distribution of plants [In Russian; Achievements Mod. Biol. (Uspekhi Sovremennoy Biologii), Vol. 7, No. 2].

Chapter IV

THE ORIGIN OF AREAS

Having acquainted ourselves with that center of an area, primary or secondary, from which a genus or a species started its dispersal resulting in the formation of its area, we may now pass to an examination of the still broader problem of the origin of the area itself. The diversity of areas of plants now inhabiting the globe and the entirely different character of the areas of many of these plants in former geological periods, as testified to by the location of their fossil remains, indicate that the origin of areas of genera and species differing in age, in ecological and biological type, and in adaptability to conditions of habitat and dispersal cannot all be explained in the same fashion. Such a mechanistic approach to this difficult problem would not in the least conform to the great diversity existing in the vegetable kingdom.

However, in order to present the problem of the origin of an area in all its entirety would mean to present the problem of species-formation in nature in all its complexity and diversity. If this were here attempted, this chapter devoted to the origin of areas would exceed in length the limit set for the entire textbook. Hence, it is quite evident that a full exposition of the problem of species-formation cannot find place in the present volume. We can merely give a very brief survey of the present status of this problem. And, first of all, we wish to emphasize the fact, proven beyond any doubt by modern science, that species arise in nature in various ways. Recent advances in biology show that the diversity of species cannot be ascribed solely to a gradual intensification of characters.

Mutations—particularly the appearance of new forms as a result of autopolyploid doubling of the chromosome number induced by the action of external factors—and, to a less extent, hybridization, accompanied by allopolyploid changes in chromosome number, played in former geological periods and still play at the present time a prominent rôle in species-formation. This does not mean that in these processes of species-formation natural selection takes no part. On the contrary, it retains all its significance, determining the survival of the fittest and their further evolution.

According to Darwin, a species arises in one definite place on the globe. From this place—thanks to a natural tendency to expand its territory, manifested, for instance, in the abundance of seeds or spores formed and in numerous adaptations of fruits and seeds enabling the future progeny to gain a foothold as far as possible from the mother plant—it begins its dispersal. It continues to expand its area in all possible directions until it encounters geomorphological or ecological barriers that it cannot cross because of purely mechanical or biological reasons. Hence, it follows: first, that the older a species, the larger the area it occupies, provided, of course, that the above-indicated barriers do not at the very outset restrict its dispersal or that subsequently there does not take place a contraction of the area previously occupied; and, second, that a discontinuous area is to be explained by the breaking up of a once-continuous area into separate sections either as a result of the dying out of the species in the intervening places, due to changes in climatic conditions, or as a result of changes in the location of seas and continents causing the rupture of such a continuous area.

Another important question needs to be clarified: Should we consider that a species arises from a single bisexual individual, two diclinous individuals, or a few individuals and that, consequently, the center of origin will occupy a territory limited by the number of progeny of this one individual or few individuals that take root, or should we consider that in the process of species-formation many individuals may participate over a considerable extent of territory, which in such case constitutes as a whole the center of origin of the area? In our opinion, both modes of origin of an area are possible, depending on the way in which a species arises.

If a species arises, for example, as a result of hybridization, the initial number of individuals of this species is determined by the number of seeds in the fruits of the given plant which succeed in finding favorable conditions for germination and growth and in maintaining themselves in competition with vegetation already established. Since plants spread by means of their dispersal mechanisms only very gradually. it is clear that the initial center of such a hybrid species will occupy a very small territory. If, on the other hand, a species arises, let us say, as a result of an autopolyploid change in the biological and morphological characteristics of an initial species caused by an increase in chromosome number induced by climatic or other conditions, the changes in such a case may affect a large number of individuals over an extensive territory, which will thus constitute the center of the area. The latter in such a case might rather be called the region of origin, but the sense remains precisely the same, since this is the initial starting point of the dispersal of a new species and the formation of its area.

Let us take, for example, a species extending its area of distribution toward the north. There comes a time when the vanguard plants attain the climatic barrier beyond which they cannot pass. If under the influence of the new climatic conditions there should arise by mutation an autopolyploid species (and that new species do so arise has now been proved for an ever larger and larger number), such a process of species-formation might embrace not a single individual plant but all the plants that had attained the indicated barrier over quite an extensive territory. Likewise, at a time of great climatic changes on the earth's surface, as, for instance, during the glacial and interglacial periods, when seas advanced or receded and great mountain masses were uplifted, the effect of such climatic revolutions also must have extended over vast areas, inducing processes of species-formation in a large number of individuals of one and the same species.

If one studies the areas of genera both as at present constituted and particularly if one adds to the present territory the regions embraced by these areas in former geological periods, he cannot but marvel at their enormous size. Such areas often occupied several continents, separated now by the waters of great oceans. Likewise a number of species (sometimes known to us only in a fossil state) of ancient genera, particularly of pteridophytes and gymnosperms but also of many angiosperms, had at one time exceptionally large areas. The remnants of these formerly extensive areas retained in the present-day flora constitute a proof that, as regards these genera and species, there took place a contraction of their areas. But this by no means signifies that such a contraction of area is an invariable rule. On the contrary, new species or species with a wide range of adaptability, finding conditions favorable for them, sometimes extend their area with startling rapidity, within a few decades becoming practically cosmopolitan.

To illustrate the process of formation and evolution of the areas of species, let us take the Angiosperms. There arose in definite centers in the Cretaceous period—and in the case of many genera probably even earlier, in the Jurassic period—genera of this group of plants, represented at that time by as yet only slightly differentiated species (or generic types, as ENGLER calls them), which attained exceptionally extensive areas of distribution. Nevertheless, we cannot assume that the flora of the entire globe was at that time homogeneous, as some authors have assumed. Climatic zones have always existed on the globe and have always served as barriers to the unlimited distribution of species. If, despite these barriers, species attained such enormous areas, this was due to the different arrangement of the climatic zones, for precisely in the Jurassic and Cretaceous periods and beginning of the Tertiary period the tropical and subtropical zones spread widely over the continents, embracing all of Europe up to its present arctic limits and considerable portions of Asia and America. Hence, it is comprehensible—assuming, in addition, a connection at that time between the continents of America and Eurasia—how the indicated genera could attain precisely on these parts of the continents such an extensive distribution. This circumstance, and also the seeming rapidity with which these plants spread, apparently facilitated the mass extinction of the Gymnosperms and Pteridophytes as a result of changing climatic conditions, which enabled the new representatives of the plant kingdom to extend their area of distribution without competition from the former inhabitants of the earth's surface.

The homogeneity of the climate over extensive portions of the continents explains likewise the slight extent to which these ancient species were differentiated. With the shifting to the south in the Northern Hemisphere of the climatic zones, with the sharp demarcation of tropical, temperate, and arctic climates, with the ever-decreasing humidity and progressive development of arid and semi-arid conditions, there developed in the second half of the Tertiary period in those regions where there had formerly existed homogeneous ecological conditions a sharp climatic zonation. Simultaneously there occurred a breaking up of these generic types, as a result of divergence, first into large groups of species, now known as collective species or coenospecies, still occupying quite extensive areas, and then into a series of small, vicarious species. This process of differentiation is still in progress, as is shown by the fact that the more our methods of research become refined, the more do we reveal small, ecologically more specialized species, formerly included in broader categories.

This process of the breaking up of collective species does not, however, show that a contraction of initial areas is the natural direction of development of present-day areas. Such an unceasing retrogression would not be in accord with the evolutionary development of organisms and actually does not take place, since the newly arisen vicarious species, from the centers of their origin, initiate anew a progressive extension of range. Coming upon favorable conditions, such species create secondary centers of development of the genus. As a result of the destruction of vegetation over considerable territories, as during the Ice Age, there penetrated, subsequently, into these territories numerous species, the areas of which reached enormous size. In many cases these species have not yet attained their limit of possible distribution.

Periods of great climatic changes have occurred on our globe many times (the above-mentioned uniformity of climatic conditions in the northern hemisphere in former geological epochs did not always prevail), and during such earlier periods of climatic changes the differentiation of genera and species must have been just as markedly expressed as at present. Subsequently, with the coming of more uniform climatic conditions, small species made place for large species, which embraced ever larger and larger territories during the course of their dispersal. These ancient extensive areas, with the disappearance of uniform climatic conditions, again entered into a period of contraction, partly as a result of the dying out of species over a portion of their areas and partly as a result of the breaking up of these areas into a number of smaller areas of new, vicarious species. These vicarious species, in their turn, begin to extend their range. As the territory they occupy increases, they become differentiated into subspecies, which may subsequently be converted into independent species. Thus, there is steadily going on a process of contraction and breaking up of old areas and the appearance in their place of new areas, which undergo the same processes of development and disintegration, making way, in their turn, to yet new species. The evolution of areas proceeds unceasingly, in conformity with the uninterrupted process of differentiation, specialization, and evolution of species.

Only the areas of relic species steadily decrease in size, as these for the most part decadent species gradually become extinct.

The theory of the origin of areas advanced by PACHOSKY (1910, 1925) may, in its basic features, be accepted, though, in our opinion, a different explanation should be given to the data presented by him. PACHOSKY, who is an opponent of the migrational origin of an area, considers that the ability to change in a certain direction is inherent to an entire genetic group, that it is equally inherent to all its representatives throughout its entire area, and that it manifests itself not in a single locality of the latter but throughout its entire territory. Hence, a race, long before it becomes an established independent unit, already occupies a definite area, this area being the result of *in situ* changes and not of migration. Such a process of area-formation PACHOSKY calls "pantopic".

To the foregoing we must take exception to the extent that we consider that such an appearance of a new geographical race or a new species is not something foreordained in the very nature of the development of a species, as might seem from PACHOSKY. It may take place, as we have pointed out above, either as a result of great climatic changes (and only then does a new race or species actually arise *in situ*) or, and this is more common under present-day conditions, as a result of the initial species during its dispersal coming upon habitat conditions, climatic or edaphic, to which it is unaccustomed and which induce its modification. Hence, the migration theory of the origin of an area preserves all its significance, although it should be noted that the term migration is often misused, an exaggerated and often indemonstrable significance being ascribed to it.

We consider, moreover, on the grounds above outlined, that an area may originate not only simultaneously over a large territory, as PACHOSKY assumes, but also within the limits of definite centers. We wish also to make clear that, even in speaking about that territory within whose limits the process of species-formation embraces all individuals of the initial species, we have in mind not the entire area but only that part of it which came under the influence of unusual (for the initial species) ecological conditions, unless, of course, the entire area came under such influence.

Likewise we may explain the tendency of whole taxonomic groups to break up into various races or species differing in color (TALIEV, 1915) as a result of the geographical isolation of different forms of a polychromic species. Isolation plays a very important rôle in the origin of new forms or new species. This is particularly evident in the case of the uplifting of new mountain chains, the separation of islands from the mainland, etc. KOMAROV (1930) has thus formulated it: "The breaking up of the earth's surface into isolated patches of dry land increases the number of species." The same idea has been expressed in a recent paper by SKOTTSBERG (1938), who points out that, when a mainland breaks up into an archipelago or group of islands, the population of any given species on such islands breaks up into a number of vicarious species (see Chapter V). Certain local micro-conditions—soil, relief, etc.—may also isolate plants from one another and serve to induce the appearance of new forms.

But if we may accept, with the reservations outlined above, PACHOSKY's area theory, this is entirely impossible as regards the so-called theory of hologenesis (ROSA, 1931). According to this theory, the evolution of a species and the direction of such evolution does not depend on the influence of environmental conditions but only on internal causes. At a certain moment, suddenly and throughout the entire territory occupied by a species, even if it be distributed over the entire globe, the maternal species breaks up simultaneously in all localities (regardless of different habitat conditions) into the same two daughter species; each of the latter, in its turn, throughout the entire territory occupied by them, breaks up into two species, etc. Thus, the evolution of organisms proceeds by dichotomous division.

Assuming uniform habitat conditions in ancient geological times, ROSA considers that the very first, most primitive species had an area embracing almost the entire globe. For a very prolonged period the species arising from this primitive species by the indicated process of dichotomy also had cosmopolitan areas. Subsequently, there began a differentiation of species, the extinction of the unfit, and the localization of the surviving species. There was thus created the present-day geographical distribution of organisms and their division into the plant and animal kingdoms.

According to this theory, there are no centers of origin of species, and areas are not formed by dispersal from these centers. A species under the influence of internal causes arises simultaneously throughout the entire area of the maternal species, which breaks up into daughter species, regardless of whether the climate is hot or cold, humid or arid, or whether or not there are present mountains or valleys, oceans, seas or rivers. There is no need to study the migrations of species; discontinuous areas are explained simply, without need of assuming connections between the continents. ROSA emphasizes that he does not deny the existence of migrations, but he asserts that they have no effect on the basic factors of the geographical distribution of species, which would remain the same even if such migrations had not taken place. In the light of this theory of hologenesis everything is apparently very much simplified. All complicated speculations as to the distribution of organisms are done away with, and at the same time biogeography as a science is done away with. If we should accept this author's viewpoint, we should have to draw our present treatise to an abrupt close.

Starting out from entirely correct paleontological data as to the extensive areas of species of former geological periods and their subsequent contraction, ROSA derives his law of hologenesis, according to which the area of a species is larger, the nearer it stands to the initial species. But in this contraction of areas he incorrectly sees a process of the gradual extinction of life on the globe, a "progressive reduction of variation", as he expresses it (ROSA, 1903). We are not in accord with this line of reasoning, since his entire theory is based on a complete disregard of the differentiation of climatic conditions on the earth, resulting in a corresponding differentiation and localization of areas of species. It does not take into account the existence, in addition to retrogressive changes, of the progressive polymorphism of the young species in present-day floras, of their expanding, not contracting, areas. We are not in accord with this theory also because of its conception of the development of organisms as a result solely of internal causes and because of its scheme of a dichotomous genealogical tree of species, which is in contradiction to the principles of evolution and to the data of modern science which confirm that scheme of the development of organisms given by DARWIN in his "Origin of Species."

Monotopic or Polytopic Origin of Areas: — Before we close our chapter on the origin of areas, we must take up one more difficult and involved problem, that as to the monotopic or polytopic origin of species and areas. By monotopic origin is meant that a species originated in a single center, from which it subsequently spread over the territory of its present area. There has also been advanced the opposite point of view, *i.e.*, that a species may originate polytopically.

Just as we can imagine the origin of identical forms and varieties at different points of the area of a species, we can also imagine the simultaneous origin of identical species at different—often entirely separate and very distant from one another—points of the area of a genus, provided that: (*1*) these points are within the area of the maternal form; (*2*) habitat conditions at these points are similar but not necessarily identical (BRIQUET, 1901).

Before proceeding to a discussion of these theories it is necessary to point out that the terms monotopic and polytopic should not be confused with the terms monophyletic and polyphyletic. The former refer to the geographical location of the center of origin of a genus or species and the origin of its area; the latter to the origin of a given taxonomic unit from one or several initial roots. In the latter case, due to adaptation to similar habitat conditions or similar biological factors, separate organs (or the entire organism) may acquire such a similar structure that there are apparent grounds for referring such similar species to the same taxonomic unit. But, naturally, once its polyphyletic origin has been established, such an artificial unit must be divided into as many separate units as initial roots participated in the development of the externally similar organisms. For instance, the subclass *Sympetalae* is of polyphyletic origin, and it should, therefore, be abolished, and the genera comprised within it should be referred to corresponding families of apopetalous *Angiospermae*. The smaller a taxonomic unit, the easier it is to trace its origin. Thus, polyphylesis of genera occurs very rarely, and, as for species, their polyphyletic origin is exceedingly unlikely. Even species of hybrid origin cannot be regarded as of polyphyletic origin, as is clear from the universally accepted definition of this term as given by us above. Such species may be called polymorphic (ROZANOVA, 1938) but by no means polyphyletic. Hence, in discussing the monotopic or polytopic origin of species, we have in mind only monophyletic species.

In order to make clearer the concept of the monotopic or polytopic origin of species, we shall cite a number of examples. Let us imagine a species distributed over a considerable territory at the foot of a mountain chain having a number of separate peaks. If this species at a certain time in its dispersal begins to ascend the slopes of this chain and, after attaining its summits, for some reason dies away on the slopes and at the foot of the chain, the habitats on the summits of the chain become absolutely isolated. Naturally, knowing the history of the dispersal of the species, we cannot regard these stations on separate peaks as having arisen polytopically.

Now let us imagine another case: A lowland species, upon reaching these same summits and being subjected to the ecological conditions of the high-mountain belt, forms a tetraploid race, which can be ranked as a separate species. All of the peaks of our mountain chain will be inhabited by this same tetraploid species, whose origin will again be monotopic, not polytopic. It arose from a diploid species having a continuous area and, if its habitats are isolated on separate mountain peaks, the reason for this lies not in the origin of the species but in the fact that the peaks are isolated from one another. If the mountain chain had been topped not by separate peaks but by a continuous plateau, the tetraploid species would have had a continuous area.

Thus, in such cases as the above, either the discontinuity of the area is only apparent, each link being actually connected with the area of the initial species, or—in case the latter has become extinct or there have occurred some great changes in the earth's surface, *e.g.*, the breaking away of part of the mainland in the form of an island—the species, though at present isolated, had their origin in the continuous area of an initial form. Hence, in order to establish the history of the flora of a given locality having in its composition isolated elements, it is necessary to penetrate deeply into the history of their initial forms and seek to ascertain whether or not their area might possibly have been continuous in the past. In such a case the work should be conducted along the lines customary for investigations in the field of historical plant geography.

Quite different will be our approach in cases where we assume the possibility of the origin of identical races in two distant and isolated points, without their occurring in intermediate localities characterized by analogous habitat conditions. For instance, WETTERHAN (1872), one of the first to advance this theory, considered it possible to assume that, as a result of slow changes in the structure of a plant and the preservation of the most favorable of these changes, there might arise identical species in entirely different and far distant localities, as, for example, in arctic and alpine regions. This same viewpoint is held by a number of other authors (see BRIQUET, 1891, 1901, and the appended bibliographies).

For example, BRIQUET (1901), in his investigation of the origin of the Alpine flora of Corsica, points out that the mountains of this island are considerably more ancient than the Alps and never were in contact with the latter. Nevertheless, on the mountains of Corsica and on the Alps there are found exactly identical species. At the same time, the species most common in the Alps and having special dispersal mechanisms for the transport of their seeds are not found in Corsica, and, conversely, the species most common on this island are not found in the Alps. On the other hand, rare Alpine species, whose transport there are no grounds at all to suspect, are found on the mountains of Corsica. BRIQUET considers that only one supposition is possible, *viz.*, that these identical species arose in such widely separated stations polytopically, *i.e.*, independently of each other.

Another example: *Primula farinosa*, which is distributed throughout the entire northern hemisphere, has an isolated portion of its area in the southern hemisphere, in South America on the shores of the Straits of Magellan, separated from the North American part of its area by a distance of ninety degrees of latitude. This antarctic primula, although it is ranked as a special variety, var. *magellanica*, is almost indistinguishable from typical *P. farinosa*. According to the adherents of the polytopic theory, the finding of this primrose along the shores of the Straits of Magellan is ascribed to its entirely independent origin there. But the question immediately comes to the fore: From what initial forms did this species with its double origin arise? Since there was not a common initial form with a continuous area of distribution, it would be necessary to assume that the forms which gave origin to this species are the initial forms not only for the genus

Primula but for the entire family of *Primulaceae*, the development of which proceeded in different directions, leading in the end to the polyphyletic origin, in two hemispheres of the globe, of entirely identical species, a conclusion scarcely acceptable to anyone.

This problem may be solved considerably more simply by assuming the very probable migration of *P. farinosa* from North America along the Cordilleras and Andes to the Far South. This is confirmed by the fact that this primula is not the only plant in South America with such a range. Here are found species of a number of genera, such as *Draba*, *Saxifraga*, *Gentiana*, *Alopecurus*, *Carex*, *Phleum*, etc., identical or very close to arctic-alpine species of these genera. According to PAX (1889) in his monograph on the genus *Primula*, *P. farinosa* first occurred in the southern hemisphere only after it had become widely distributed in the northern hemisphere. Its migration to the south may have taken place during the Ice Age, when the climatic conditions in the Andes were more humid than at present, which favored the growth of arctic-alpine species. In the post-glacial period these species were preserved in the mountains of North America and the antarctic region of South America but died out in the intermediate habitats, a bipolar area thus being formed.

Similarly, as regards the high-mountain vegetation of the islands of the Tyrrhenian Sea, which served as the basis for the conclusions drawn by BRIQUET, we find in the work of BRAUN-BLANQUET (1923) on the origin of the flora of the central massif of France the following: "The hypothesis . . . of a 'polytopic' origin . . . gradually vanishes in thin air and has to yield its place to a better grounded explanation that is also in accord with morphogenetic data. The latter show that the definitive separation between the Betic Cordillera and the Moroccan Rif took place at the beginning of the Pleiocene period; the islands of the Tyrrhenian Sea were separated from the mainland at the end of the Tertiary period, at which time they already possessed a diverse, orophytic flora. On the summits . . . of the Central Massif there must also have existed alpine species" (pp. 206–7).

The advocates of the polytopic theory consider possible the independent origin by mutation of absolutely identical species in two, widely separated points of the initial area. But up to the present time we do not know of a single case of the origin by mutation of one and the same species, with an absolutely identical complex of characters, at widely separated points. Likewise there cannot arise one and the same hybrid species at two independent and very distant points of the areas of the parental species. Species having extensive areas are not fully homogeneous throughout the entire extent of their areas, but constitute series of vicarious races. Consequently, when these species are crossed at widely separated points of their areas, the differences between the geographical races to which the parental forms belong exclude the possibility of the occurrence of completely identical hybrid species.

All this leads us to the conclusion that similar habitat conditions may induce the polytopic origin of similar forms, distinguished from one another by only a few characters, *e.g.*, glabrous or pubescent forms, forms with glandular hairs, with liguleless leaves (in cereals), with or without anthocyanin, etc. Such parallel forms may arise in different parts of the area of a species either by mutation or in consequence of hereditary causes, due to the parallel variation of the characters of allied forms, which "are apt to vary under similar exciting causes in a similar manner" (DARWIN, 1911, p. 585). Such parallel variation, called by DARWIN "analogical variation", was later, on the basis of extensive material, formulated by VAVILOV as the "law of homologous series in variation". But if this may be accepted as regards separate characters, changes in which may be caused by separate factors, there is still no basis for assuming the possibility of the polytopic origin of species, characterized by a whole complex of characters, since no two localities on earth are exactly alike in all respects, as regards all factors, all physico-geographical conditions. In any case, the advocates of the polytopic theory have presented only theoretical assumptions and have not advanced a single real proof. Even when the discontinuity of an area seems inexplicable, we may in most cases expect that geological data will be discovered, if they have not already been discovered, which will explain the configuration of such areas on the basis of purely historical causes.

References:

BRAUN-BLANQUET, J., 1923: L'origine et le développement des flores dans le massif central de France (Paris-Zürich).

BRIQUET, J., 1891: Recherches sur la flore du district Savoisien et du district Jurassique franco-suisse (Engl. Bot. Jahrb., Vol. 13).

BRIQUET, J., 1901: Recherches sur la flore des montagnes de la Corse et ses origines (Ann. Conserv. et Jard. Bot. Genève, Vol. 5).

BRIQUET, J. 1905: Le développement des flores dans les Alpes occidentales (Rec. scient. Cong. intern. bot., Vienna).

KOMAROV, V. L., 1930: The genus *Phacellanthus* S. & Z. (*Orobanchaceae*) in the Far East (In Russian; Bull. Acad. Sci. U.R.S.S. No. 3).

LEAVITT, R. G., 1907: The geographic distribution of closely related species (Amer. Nat., Vol. 41).

LECLERCQ, S., 1932: Paléontologie et migration (Extr. du C. R. du Congrès Géogr. de Bruxelles, Section de Biogéographie).

PACHOSKY, I., 1910: Chief Features in the Development of the Flora of Southwestern Russia (In Russian).

PACHOSKY, I., 1925: Area and its origin (In Russian; Jour. Russ. Bot. Soc., Vol. 10, No. 1–2).

PAX, F., 1889: Monographische Übersicht über die Arten der Gattung *Primula* (Engl. Bot. Jahrb., Vol. 10).

ROSA, D., 1903: Die Progressive Reduktion der Variabilität (Jena).

ROSA, D., 1931: L'ologénèse (Paris).

ROZANOVA, M. A., 1938: On the polymorphic type of origin of species (Comptes Rendus (Doklady) Acad. Sci. U.R.S.S., Vol. 18, No. 9).

SCHWARZ, O. 1936: Über die geographisch-morphologische Methode in der systematischen Botanik (Mitt. deutsch. dendr. Ges., Vol. 48).

SKOTTSBERG, C., 1938: Geographical isolation as a factor in species formation (Proc. Linn. Soc. London, 1937–38, 150 Sess., Part 4).

TALIEV, V. I., 1915: An Attempt to Investigate the Process of Species-Formation in Nature, I (Kharkov).

WETTERHAN, 1872: Über die allgemeinen Gesichtspunkte der Pflanzengeographie (Ber. Senckenb. Naturf. Ges. Frankf. a/M.).

Chapter V

TYPES OF AREAS

The areas of species of plants are as diverse as the distribution of plants is varied. It is, consequently, impossible to group all this diversity into a definite number of types. Any hard and fast system of classifying areas would be artificial, since it would not reflect the natural diversity of plant distribution. Nevertheless, a definite terminology for designating different types of areas is necessary. Attempts to elaborate such a terminology have been made by various investigators. We give below two examples.

According to PACHOSKY (1921, p. 209), the following types of areas may be distinguished: (1) two races or two closely allied species may inhabit one and the same region, *i.e.*, may grow side by side—a *coextensive area;* if the area of one such species is smaller than that of the other, the smaller is called an *included area* (BUSH, 1917); (2) two species may inhabit different regions, in which case their areas are either *separate* or *contiguous;* (3) the areas of two species, at the point of contact, may overlap—an *overlapping area;* (4) the area of distribution of a species may be broken by the area of another species and then again be resumed (such an alternation of species may be repeated several times over)—a *discontinuous area;* (5) within the area of distribution of a species there may be enclosed islet-like habitats of another species, which, in addition, has its own main area, that may or may not contain islet-like inclusions of the first species; (6) lastly, there are cases when one of the species has an independent area, while the other only has such islet-like habitats within the area of the first.

ILYINSKY (1933) proposed the following classification of areas, which takes into account not only their statics but also their dynamics:

I. *Progressive or expanding areas*
   1. Radiate area (*area radiata*)
   2. Fringed area (*area fimbriata*)
   3. Continuous area (*area solida*)

II. *Retrogressive or contracting areas*
   4. Reticulate or perforated area (*area perforata*)
   5. Discontinuous or fragmented area (*area disjuncta* or *fragmentata*)
   6. Area limited to one greatly restricted geographical region (*area solitaria*)

The areas of species distributed both in the northern and southern hemispheres but not in the tropical zone are called *bipolar areas.*

Other types of areas, such as cosmopolitan, endemic, vicarious, and relic areas, as well as continuous and discontinuous areas, we shall discuss at greater length.

Cosmopolitan and Endemic Areas: — There is only a very limited number of species that are distributed over all or almost all the globe. Species so distributed are, for the most part, those which are indifferent to environmental conditions, *i.e.*, ubiquitous species. Such wide-ranging species are known as *cosmopolites* or *pan-endemics* (FENZL).

In reality, there are no true cosmopolites, *i.e.*, species inhabiting the entire globe from pole to pole, at least not among the higher plants. As ALPHONSE DE CANDOLLE (1855) pointed out, such species cannot exist. There are species distributed from the arctic through the temperate zone and then reappearing in the southern hemisphere; there are also species distributed throughout all the tropics and even considerably beyond them. But there are no species, at least no low elevation species, distributed on the equator and at the same time reaching nearly to the two opposite poles. This is due to the great differences in climatic and edaphic conditions, and holds true both for species naturally distributed and for those which have accompanied man. Hence, cosmopolitanism, in the absolute sense of this word, does

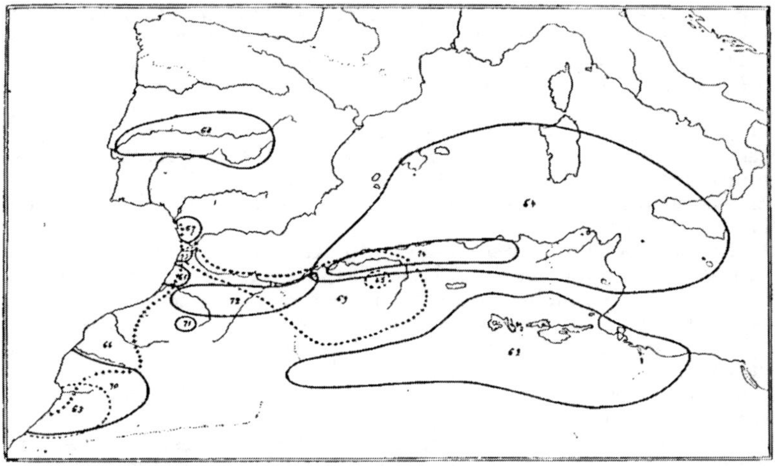

FIG. 6. — Distribution of species of *Celsia* in southern Europe and northern Africa: 61, *C. ramosissima;* 62, *C. Ballii;* 63, *C. longirostris;* 64, *C. cretica;* 65, *C. pinnatisecta;* 66, *C. lyrata;* 67, *C. laciniata;* 68, *C. Barnadesii;* 69, *C. Battandieri;* 70, *C. maroccana;* 71, *C. zaiamensis;* 72, *C. Faurei;* 73, *C. commixta;* 74, *C. betonicifolia.* (After MURBECK). — 65, 74 included areas; 63, 64, 69, 70, 72 overlapping areas; 62, 68, 71 separate areas; 61, 67, 73 contiguous areas.

not exist, and if we use this term, it is only in a relative and approximate sense. The number of species occupying as much as half of the earth's surface is very limited and does not exceed 20 or 30 species. DE CANDOLLE could enumerate only 18 such species, and even if we assume that closer acquaintance with the distribution of plants would show the inaccuracy of this figure and that it should be doubled or trebled, the number of such species is nevertheless insignificant. Even the total number of species occupying one-quarter of the earth's surface is insignificant in comparison with the total number of species. According to DE CANDOLLE, it is not over 200 (to be accurate—116), which amounts to only 0.001 per cent of the total number of flowering plants known in his time.

It is of interest that among the species in this list are plants differing widely in habitat. There are fresh-water aquatics (15 floating or submerged species and 23 species that root in the bottom of ponds or lakes but extend the upper part of their stems and leaves above the surface of the water) and, on the other hand, there are xerophytes (14–15 species). Another 30 or so species are accounted for by weeds infesting cultivated fields. The life span of these species also differs: 47 of them are annuals, 3 biennials, and 66 perennials. Likewise they are distributed at random among the various groups of the vegetable kingdom.

For these wide-ranging species MOLISCH (1926) has proposed the more accurate term "semi-cosmopolites".

If we turn now to species not occupying such a large territory on the earth's surface, we can note, as regards their distribution, two main types. To the first belong species distributed in various regions and countries; these are known as *scattered*, *sporadic* (DE CANDOLLE), or *polyendemic* (FENZL) species. In contrast to these, there are other species which in their distribution are limited to a very restricted area, not extending beyond some one region, country, island, or mountain summit. Such species are called *endemics*. With respect to all these terms, just as in the case of the term "cosmopolite", it must be remembered that they bear only a relative, often conditional, and by no means absolute character.

By *endemic area*, a concept first introduced into science by A. P. DE CANDOLLE, we now understand the area of a taxonomic unit, particularly of a species, limited in its distribution to some one natural region or habitat, the physico-geographical and ecological conditions of which set it off from adjoining regions. In this respect islands and mountain peaks, with their distinct natural boundaries, make possible the most exact determination of an endemic area. As regards other regions, such a sharp delimitation rarely occurs.

But the very concept "endemic" is not simple. There are different interpretations and evaluations which depend largely on the age of the endemic. In some cases the origin of an endemic should be referred to remote times, judging by its phylogenetic antiquity, taxonomic isolation, character of its habitats (*e.g.*, mountain peaks or islands, which have since time immemorial been in a state of isolation), or its past history. For example, to this group we may refer a species which at one time had an extensive distribution but which later, during the course of geological revolutions accompanied by climatic and physico-geographical changes in habitat conditions, gradually died out and eventually entirely disappeared from the greater part of its area, surviving only in a small portion of the latter, thanks to some favorable circumstances or other. Consequently, in ancient lands, as, for instance, in mountain regions or on islands the vegetation of which has survived during geological revolutions, the percentage of endemic species is necessarily very high, and the endemism itself is of an ancient, relic character. Hence, the number of relic endemics in the composition of a given flora indicates both how old the latter is and how long it has been isolated. The Alps, for instance, have 200 endemic species, the Canary Islands—469 species (45% of the total),

Corsica—58 per cent, Madagascar—66 per cent, New Zealand—72 per cent (1,000 species), Hawaii—82 per cent, and St. Helena—85 per cent.

In other cases, however, endemic species—or, more frequently, units of a lower order, such as subspecies, varieties, forms—are characteristic of younger portions of the earth's surface. Then the endemism has an entirely different, more recent origin. This may be the case, for instance, when within the limits of some natural region, due to changes in ecological conditions, an initial form gives rise to new forms. The latter may be closely bound, due to certain specific habitat conditions, to this region only, or for some other reason they may remain within the given region and not spread beyond its confines.

A similar origin of endemics of this type is observed also in case of the relatively recent isolation of a given region. A very striking example of this is the Crimean Peninsula, which at one time constituted part of the eastern Mediterranean region, being connected as late as the Pleiocene period with Asia Minor and Transcaucasia, but which later separated from them and became attached to the South Russian mainland. Because of the fact, however, that the isthmus by which it is attached is still characterized by solonchak soils unfit for the growth of most plants, the Crimean Peninsula is, in a biological sense, almost an island, having at a comparatively recent date entered into a condition of isolation. Consequently, in the Crimea ancient, relic endemism is represented by only a few species, the number of which, as further studies are made of the flora of Asia Minor, becomes less and less, while secondary endemism, of more recent origin, becomes more and more extensive, embracing ever new forms and species, as we study the flora of the Crimea more thoroughly. And this holds true not only as regards the flora but also as regards the fauna of the peninsula (Wulff, 1926).

As a similar example we may take Cape Colony, which, as regards climatic conditions, is entirely isolated from the adjoining parts of Africa and, hence, very rich in endemics of comparatively recent origin. The same holds true in Western Australia, which is cut off from the other parts of Australia by a desert zone.

Such data forced Engler (1882, Vol. II, p. 48) to point out that "it should never be forgotten that there are two kinds of endemism—one based on the preservation of ancient forms, which may have originated in entirely different regions, and the other based on the development of new, entirely autochthonous forms". Likewise Braun-Blanquet (1923, p. 223) is correct when he states that: "The study and precise interpretation of the endemism of a territory constitute the supreme criterion, indispensable for arriving at any conclusions regarding the origin and age of its plant population. It enables us better to understand the past and the transformations that have taken place; it also provides us with a means of evaluating the extent of these transformations, the approximate epoch when they occurred, and the effects which they produced on the development of the flora and the vegetation".

For these two kinds of endemism different authors have adopted different names. Drude (1890) calls them *relic* and *secondary* endemism; BRIQUET (1905)—"endémisme par conservation" and "endémisme par novation"; DIELS (1908)—*conservative* and *progressive* endemism; HERZOG (1926)—*relic* or *ancient* endemism (Altersendemismus) and *neo-endemism;* BRAUN–BLANQUET (1923) and CHEVALIER and GUÉNOT (1925)—*paleo-endemism* and *neo-endemism.* The last-mentioned terms seem to us the most suitable, and we have decided to adopt them, despite the fact that, as RIDLEY (1923, 1925) has pointed out, some confusion arises from the use of one and the same term "endemic" to designate two different types of plants. Though both are characterized by a restricted area of distribution, one type embraces relics of earlier floras which have survived in a limited portion of their past territory, the other type species having originated in a given region and not yet having spread beyond it. In view of the desirability of two distinct terms by which to designate these two types, RIDLEY proposes to reserve for the second type the term "endemics", since they arose from parental forms on the territory of their habitat and continue to live, so to say, among their own "demos" (from which Greek word "endemic" is derived). In contrast to this plants of the first type, being remnants or survivors of former floras, he calls *epibiotics*. However, if the latter term is used, it seems to us advisable to restrict its use to endemic relics (*i.e.*, paleo-endemics) and not to apply it to relics in general.

To distinguish between these two types ordinarily does not present any particular difficulty, since paleo-endemics (or epibiotics) do not have any close connection with other species in their area, whereas neo-endemics have numerous, often close bonds with other species in their region of distribution. However, there are endemics which it may be difficult to refer to either of these categories. For example, there are instances of endemic genera, which without any doubt should be classed as paleo-endemics but which comprise species that should be classed as neo-endemics. These RIKLI calls *active epibiotics*.

The endemic flora of Australia, according to DIELS (1906), constitutes another example of the difficulty of referring endemics to the two indicated categories. He divides this flora into three groups: (*1*) endemic genera not having any connection with pan-Australian genera; often monotypical genera; (*2*) endemic genera having some connection with the indicated genera; (*3*) endemic genera having undoubted connection with pan-Australian genera.

Endemic species are sometimes confined to exceptionally limited areas, *e.g.*, to a single mountain peak. Such species may be called *local endemics*. Their greatly restricted areas may be due to one of three causes: (*1*) the recent origin of the endemic, which has only begun its dispersal and the formation of its area; (*2*) the antiquity of the endemic, which is contracting its area, in consequence of which the limited territory now occupied by it constitutes the last remnant of its former area; (*3*) the specificity of the habitat conditions, which prevail only in the given spot, or the impossibility of an expansion in area due to obstacles of a physico-geographical nature.

BRAUN–BLANQUET (1923) distinguishes, in addition, *micro-endemics*, which are endemic forms of lower taxonomic rank and recent origin.

Lastly, we wish to mention *pseudo-endemics*, to which HERZOG

(1926) refers those species that have been encountered only once and only in one place and that are, for the most part, mutants, being likely to disappear as suddenly as they arose.

Besides species whose endemism is the result of the history of their origin and distribution there are species whose endemism is connected with definite habitat conditions. The latter may be called *ecological endemics*. As is well known, there are a large number of endemic species which arose as a result of edaphic conditions, *e.g.*, species connected with cretaceous soils, such as *Linaria cretacea*, *Silene cretacea*, *Scrophularia cretacea*, and *Hedysarum cretaceum;* with sandy soils—*psammo-endemics*, in the terminology of LAVRENKO (1936)—such as *Agropyrum tanaiticum* (in the Don Valley) and species of the series *Centaurea margaritacea s.l.* (in the Don and Dnieper Valleys); with granite, serpentine, and solonchak soils. Moisture conditions are likewise a cause of the occurrence of endemic species; among such species we may mention those that grow along the banks of rivers (FURSAEV, 1937).

The determination of the type of any given cases of endemism, particularly the division of endemic species into paleo- and neo-endemics, is a very important factor in the analysis of a flora. In retracing the history of a flora paleo-endemics may serve as important guideposts. The paleo-endemic nature of a species is usually determined by its relic character, its geographic and often taxonomic isolation, the absence of variation in characters, its narrow restriction to definite ecological conditions, upon all of which depend the limited size of its area and, to a great extent, its retrogressive nature. But all these factors are not sufficiently objective to determine positively the age of an endemic. For this purpose the use of cytological data has been proposed, *e.g.*, for the genus *Festuca* (LEWITSKY and KUZMINA, 1927) and, more recently, for the genus *Agrostis* (SOKOLOVSKAYA, 1937). Taking into consideration the circumstance that, when a genus comprises a polyploid series of species, those species having a small chromosome number are older than and constitute the initial forms for the species having the larger chromosome numbers, we may use this character in determining the relative age of any given case of endemism. Endemic species with a small chromosome number may usually be referred to paleo-endemics, while neo-endemics have larger chromosome numbers. For example, the endemics *Agrostis planifolia* (a Caucasian species) and *A. hissarica* (a Central Asiatic species) have 42 as their diploid chromosome number, while the endemics *A. Biebersteinii* (a Caucasian species) and *A. Trinii* (an East Siberian species) have only 14. It is clear that the former should be regarded as neo- and the latter as paleo-endemics.

Vicarious Areas: By vicarious areas we mean areas for the most part mutually exclusive and belonging to closely related species, differing only in a few specific characters and linked by the fact that they derived their origin from one initial form. The formation of vicarious areas has at its basis the process of the genesis of geographical races. Speaking of vicarism, we have in mind chiefly species and geographical races (subspecies), although there may also be vicarious sections of genera, vicarious genera, and even, in a certain sense, vicarious families. The number of such vicarious taxonomic units, however, is insignificant as compared with the exceedingly large number of vicarious species and races. In any present-day taxonomic monograph we may find many instances of such geographical series of species (in the narrow sense) or subspecies (races), inhabiting independent geographical areas usually mutually exclusive but sometimes slightly overlapping. Examples of such vicarious areas are given in the accompanying maps.

In some cases vicarious species have arisen as a result of climatic changes, changes in the distribution of seas and land, and other causes of a historical nature. In such cases vicarious areas have an ancient character, and the species themselves are usually paleo-endemics. In other cases the origin of vicarious species or forms may have been occasioned by differences in the ecological conditions encountered by the species in the course of its dispersal.

The spread of a species from one ecological region to another, accompanied by the formation of vicarious species, may be designated (Kashkarov and Korovin, 1931; Korovin, 1934) as *autonomous migration*, as distinguished from *successive migration*, when a species during its dispersal preserves its chief characters, due to its finding itself constantly in a similar ecological environment or to its adaptability to a wider range of ecological conditions.

Soviet Central Asia, thanks to the exceptional diversity of its ecological conditions, gives numerous examples of such migration of species of different geographical and ecological origin. For instance, according to Korovin (1934), the genus *Scaligeria*, whose area has its initial center in the Mediterranean Basin proper and a secondary center in southern Turkmenistan, is represented by the following vicarious species, whose vicarism is both horizontal (geographical) and vertical (altitudinal): *S. transcaspica*—deserts; *S. hirtula*—steppes; *S. ferganensis*—woods; *S. ferganensis* var. *Korshinskyi*—subalpine meadows. The same holds true for the genus *Bunium*: *B. capusi*—deserts; *B. persicum*—semi-deserts; *B. chaerophylloides*—steppes; and *B. Angreni*—alpine zone.

Species of the genus *Phlomis* belonging to the subgenus *Phlomidopsis*, which originated in eastern Asia, are distributed in Soviet Central Asia primarily in subalpine and alpine zones. But one of the species of this subgenus, *P. brachystegia*, grows in meadow steppes, and another, *P. Popovii*, still lower in the steppes proper.

Arthrophylum haloxylon, which grows on solonchak soils, is replaced on sandy soils by *A. persicum*. *Aristida pennata*, which grows on stationary sands, is replaced on shifting sands by *A. Karelini* and on stony soils enriched with gypsum by a third species, *A. rigida*.

Pachosky (1921) likewise pointed out that corresponding to different soils and other diverse conditions there exist pairs or groups of species, which may be united in coenospecies (collective species). Moreover, these species, though linked with definite habitat conditions, are not remnants of a once-existent flora and are not genetically isolated from one another, but are, on the contrary, new forms of comparatively recent origin and still in the process of development. Such forms occur most frequently in mountainous regions, due to the diversity of habitat conditions found there.

The diverse types of vicarious areas were subjected to a critical and detailed analysis by VIERHAPPER (1919), and we shall summarize this analysis here.

The chief type of vicarism may be called *spatial vicarism*, and, depending on whether the vicarious species are distributed horizontally or vertically, we may distinguish between *horizontal* and *vertical vicarism*. The first is of more common occurrence than the second, since the latter naturally may be found only in mountainous regions. Spatial vicarism may occur in different regions isolated from or mutually

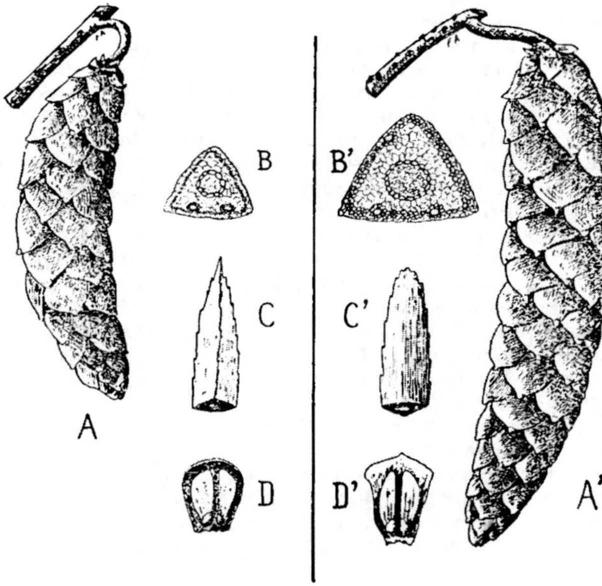

FIG. 7. — Vicarious species of pine, *Pinus Strobus* and *P. monticola*, in North America. (After VICTORIN: "montrant la différenciation morphologique due à l'isolement géographique. L'isolement est ici le résultat de l'extension, à la fin du Crétacé, d'une mer intérieure qui sépare l'Amérique du Nord, dans le sens de la longueur, en deux massifs continentaux. A–A', Cônes; B–B', Section des feuilles; C–C', Pointes des feuilles; D–D', Écailles et graines.").

exclusive of one another—*regional vicarism*. In other cases vicarious species may be found within one and the same region but under different habitat conditions—*intraregional vicarism*. Such conditions include different kinds of soil substrata, *e.g.*, volcanic rock and limestone, solonchak and podzol soils, sandy and hardpan soils, shady and sunny habitats, etc. Among vicarious forms of this type there should be mentioned the many so-called colored species or races, the color of whose flowers changes depending on local habitat conditions.

Contrasted to this spatial vicarism is vicarism linked with the *time of development* of the plant. To this latter type belong species characterized by seasonal dimorphism, as established by WETTSTEIN. By this we understand, for example, the finding on one and the same meadow two closely related forms differing in time of development, one flowering in the spring, the other in the autumn. Such seasonal dimorphism is occasioned by a change in the initial form, resulting in its conversion into two local varieties under the effect of the economic use of the meadow by man.

All forms of intraregional vicarism, including seasonal dimorphism, not having geographical significance, are only of minor interest for historical plant geography.

In addition to the various types of *true vicarism*, there is also a vicarism that is only seeming, called by VIERHAPPER *pseudo-vicarism*. While in the case of true vicarism the vicarious species arise from a common initial form, becoming differentiated either within the limits of the latter's area or upon penetration into new habitat conditions followed by isolation, in the case of pseudo-vicarism the seemingly vicarious species, although related to one another, have arisen from different initial forms, and their apparent vicarism is occasioned by the secondary penetration of a second species into the area of a first and by its occupation of those portions of this area not occupied by the first species due to their not being suited to its biological peculiarities. Usually such pseudo-vicarism may occur only in the case of the absence of truly vicarious species. Hence, in the case of true vicarism the penetration of the initial form into a given region is the first step, followed by its breaking up into vicarious races adapted to different habitat conditions. In the case of pseudo-vicarism, on the other hand, the origin of the different races is the first step, after which comes the occupation of separate territories.

Pseudo-vicarism may be of various types, *viz.*: (*1*) the two forms penetrating a given region may be fully established species or races; (*2*) one of the species or races has its origin in the given region, while the other does not penetrate this region until later. As an illustration, we may cite an example given by VIERHAPPER. The genus *Erigeron* has a number of vicarious species, geographically markedly distinct. Among them there are two species in the Alps: *Erigeron polymorphus*, growing on calcareous soils and descending far down toward the foot of the mountains, and *E. uniflorus*, in contrast to the other, avoiding calcareous soils and closely confined to the higher altitudinal zones. The nature of these two species gives apparent grounds for believing that they were derived from a common initial form and became differentiated as a result of adaptation to different, mutually exclusive habitat conditions. However, a study of the entire cycle of forms to which these two species belong has shown that we have to do here with species which, although related, are of different origin. *E. uniflorus* arose, presumably, not in the Alps but in the mountains of Asia or in the Arctics, whence it penetrated during the Ice Age into the Alps within the limits of the area of *E. polymorphus*, already established there at the time, and occupied places not already occupied by the latter because of edaphic conditions.

Pseudo-vicarism, consequently, despite its being seemingly identical with true vicarism, is a phenomenon simulating the latter but of an entirely different nature.

The problem as to the origin of vicarious species and areas has constituted one of the most difficult problems of plant geography, inadequately solved until quite recently. As early, however, as 1869 KERNER pointed out that the present area of a species cannot always be regarded as the limit of its possible distribution; in many cases its present boundaries represent only those which the species at the given moment has reached in its distribution, the explanation of its absence on the other side of these boundaries being simply that it has not yet had time to go beyond them. But, at the same time, it is perfectly clear that in many cases habitat conditions—climatic and edaphic—

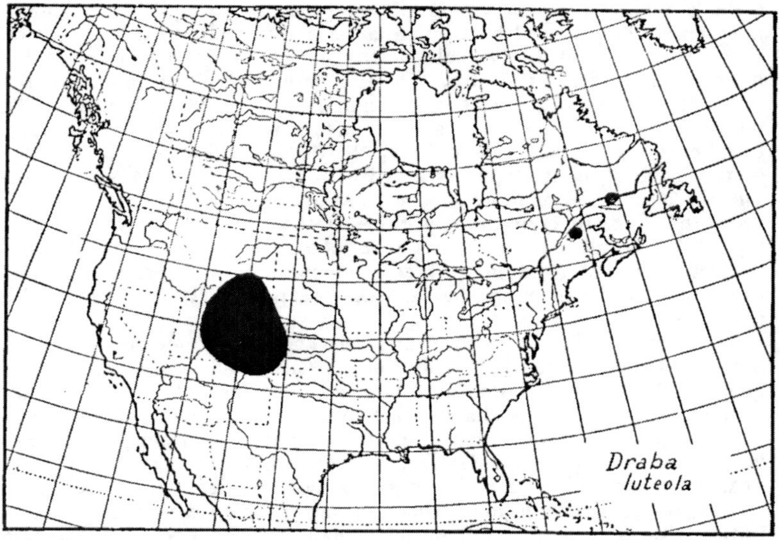

FIG. 8. — Area of distribution of *Draba luteola* in the Rocky Mts. (main area), and of its vicarious variety, var. *minganensis* (the two small black spots) near the Gulf of St. Lawrence, the latter having arisen as a result of isolation. (After VICTORIN).

constitute insuperable obstacles for the species in its distribution. Such conditions may be, for example: decrease in the sum total of heat during the summer months, shortening of the vegetative period by early spring or fall frosts, summer drought, decrease in the annual precipitation, change in the distribution of precipitation through the seasons of the year, replacement of some soils by others markedly differing in chemical or physical properties, etc. In addition, there is no doubt that on the periphery of an area, toward the margins of the range of a species, the competition of other species may also constitute a barrier to further distribution.

These circumstances compel us to conclude that the area of any species may be divided into two parts. In one the species finds itself in optimum conditions, and the entire cycle of its development is each year realized without hindrance. In the other part of the area, comprising chiefly the periphery, the habitat conditions are only to a minimum degree suited to its biology. Here the species may mature its seeds only in favorable years, while in unfavorable years, if it finds this at all possible, then only in especially sheltered spots. This leads to a decrease in the number of progeny. Hence, nearer the periphery of its area a species is represented by an ever smaller number of individuals growing in ever more rare and isolated spots, while nearer the center of its area it is represented by an ever larger number of individuals occupying an ever wider range of habitats.

In the center of its area, where habitat conditions are most favorable for the existence of a species, there is, in KERNER's opinion, very little probability of the origin of new species, since, if such should arise there, they would be eliminated as a result either of competition with the numerous individuals of the initial species or of inevitable hybridization with the latter. On the periphery of the area, on the other hand, these obstacles to the origin of new species disappear. The initial species no longer covers the territory so completely, some habitats being unsuitable or inaccessible to it; hybridization with it, due to the isolation of its habitats and to the decrease in number of individuals, becomes ever more rare or does not occur at all. At the same time, the different climatic conditions at the boundary of an area further the origin of aberrant forms even in species that are not ordinarily subject to great variation. These circumstances make clear the relative abundance of young species near the periphery of areas of old, initial species.

The periphery of the area of a species should be understood to refer not only to its horizontal distribution. Altitudinally there arise analogous (to those just described) changes in climate and soil substrata and, consequently, analogous changes in habitat conditions. Hence, even in that portion of the area where, due to favorable conditions, the species is represented by the greatest number of individuals, there may exist analogous possibilities for the origin of new species. In consequence of this, the fact that on maps there are often shown within the areas of initial species tiny areas denoting young species does not indicate intermingling, since this is only seemingly so, as they are located at different elevations above sea level.

Changes in edaphic conditions presumably led to the origin of such specialized species as *Androsace Hausmanni* and *Asplenium Seelosii*, which grow in southern Tyrol, where the soil is predominantly of dolomite formation, and which were apparently derived from *Androsace glacialis* and *Asplenium septentrionale*, which are linked in their habitats with slate soils. Similarly, the unique species *Asplenium serpentini*, which grows on serpentine soils in Austria, Moravia, and Bohemia, apparently arose from *Asplenium adiantum nigrum*, which grows on non-serpentine soils. The existence of such species, now known as vicarious but which KERNER called "parallel forms", testifies to the influence of external factors on the origin of species.

Such new species are able to extend their areas beyond the limits, set by climatic and soil conditions, to the distribution of the initial species. Hence, there arise species extending farther north or higher up in the mountains than the initial species.

The above-described phenomena do not take place in the case of relic species in which the capacity for variation is greatly lowered. Finding themselves in unfavorable environmental conditions, they cease to multiply and gradually die out, their area of distribution becoming more and more contracted, being preserved only in occasional, isolated habitats.

Even earlier than KERNER, MORITZ WAGNER (1868) advanced his law of plant migration (which we have already mentioned), based on similar observations. In this work, as well as in his later work, "Über den Einfluss der geographischen Isolierung und Kolonienbildung auf die morphologischen Veränderungen der Organismen" (1870), WAGNER developed the idea that hybridization, inevitable among individuals growing together in a habitat, results in the smoothing over of any deviations arising as a result of the variability of organisms. An individual plant with hermaphrodite flowers (or a pair of plants having diclinous flowers), happening to penetrate into a locality isolated from the region of mass distribution of representatives of the same species, undergoes, as a result of external factors, changes that are preserved due to isolation. This isolation may be expressed not only in great distances or the transfer of occasional individuals over obstacles of various kinds—mountains, seas, rivers—hindering the distribution of the initial species; it may be *any* spatial isolation, regardless of extent, provided it makes possible the preservation of changes induced by habitat conditions—a circumstance that KERNER did not take into account.

Since KERNER's and WAGNER's time, particularly during the past forty years, the external morphology of plants has ceased to be the only criterion for determining their taxonomic position. Geographical distribution has acquired significance as one of the chief criteria for establishing the existence of a taxonomic unit of the rank of a species or subspecies. Hence, it must be recognized that geographical factors exert an influence not only on the distribution but also on the origin of a species.

As a result of the adoption of the geographico-morphological method in studies of genera and species, a method that was given a firm basis by the works of WETTSTEIN (1898) and KOMAROV (1901), many data have been accumulated and critically evaluated which leave no doubt that the origin of vicarious species is linked with the effect of environmental factors encountered during the course of the dispersal of these species. This has been confirmed by the mutation theory, according to which mutations manifest themselves not only in changes in major characters but also in numerous minor changes.

In 1927 MULLER induced mutations in *Drosophila* by X-ray treatment and thus demonstrated that such changes may arise not only as a result of internal causes but also as a result of external conditions. This controverted the view, formerly held to be incontrovertible, that the rôle of the environment in the formation of new hereditary characters is not of a creative nature, that the environment cannot directly induce the appearance of new characters in plants and animals. These experiments have been repeated by many investigators, not only with animals but also with plants, and it has been definitely established that mutations may arise as a result of external factors not only under artificial conditions but also in nature.

The origin of new characters by mutation, linked with changes in the germ-cells, may be accompanied either by changes in separate hereditary factors, gene mutations, or by a change in chromosome number, a chromosome change. As regards the occurrence of gene mutations as a result of external influences, we know as yet only cases where they were artificially induced under experimental conditions. In contrast to this, the ever-increasing number of established cases of chromosome changes include some where they undoubtedly occurred in nature as a result of environmental conditions.

From an evolutionary point of view of particular interest is the fact that an increase in chromosome number may be accompanied by heritable morphological changes, thus providing a basis for giving such polyploid forms specific rank.

Polyploidy may arise as a result of an increase in the chromosome number following hybridization, *allopolyploidy* (which for problems of plant geography are of less interest), or as a result of a simple doubling of the plant's own chromosome set, *autopolyploidy.* The occurrence of autopolyploidy as a result of the effect of external factors is now established beyond any doubt. It has been induced experimentally by treating the vegetative tissues or germ-cells with various agencies (temperature, X-rays, colchicine, etc.). It may also be considered proved that the phenomenon of polyploidy, including autopolyploidy, is of common occurrence in nature, constituting an important factor in species-formation.

A study of autopolyploid species and forms has shown that in many cases they have very definite geographical areas differing from the areas of the initial diploid species. Moreover, these areas are concentrated, for the most part, in mountainous regions in the higher altitudinal zones or in far northern or equatorial latitudes, or in other localities where conditions are unusual for the given plant. This circumstance indicates that extreme climatic or other habitat conditions induce doubling of the chromosome number presumably as a result of irregularities at meiosis. Chromosome doubling is linked with the origin of a new mutant form differing from the initial species in a number of characters, which accounts for its adaptability to life under the indicated extreme habitat conditions. These conclusions are of prime importance for plant geography, since they reveal one mode of origin of vicarious species.

Chromosome doubling within the limits of a species usually occurs not throughout the whole extent of the latter's area but only in a portion of it. The new polyploid forms differ not only cytologically but always, to a greater or less degree, in their morphological structure. This morphological difference may be expressed both in qualitative and quantitative or only in quantitative changes. The taxonomic significance of these morphological races is rated variously by different investigators. Some consider that these changes are only of an intraspecific nature; others regard them as adequate basis for ranking the new forms as independent species. In a number of cases such new forms were given specific rank on the sole grounds of a morphological study and the establishment of the existence of independent areas, and it was only later ascertained that these new species were polyploid forms. In other cases the establishment of the polyploidy of forms morphologically differing but slightly from the type has explained much that was incomprehensible in their geographical distribution.

Chromosome doubling is accompanied not only by changes in morphological structure but also by changes in the biological properties of a plant, which likewise play an important rôle in determining the geographical distribution of these new species and forms. The distribution of plants, the fact that they have clearly defined areas, is determined by ecological factors now existent or prevailing during earlier periods in the history of the given species. Only a few species of plants possess such a wide range of adaptability to extreme ecological conditions as to enable them to become distributed over practically the entire globe and as to serve as grounds—although not without straining the point somewhat, as we have seen—to call them cosmopolitan. Other species possess a narrower ecological adaptability, which results in their areas of distribution being more localized. In the latter case it is clear that, if for one or another reason changes occur in the biological properties of a species, its dependence on ecological factors is likewise changed, and this causes changes in its area of distribution, which may now extend beyond those boundaries which set a limit to the spread of the initial species. The occurrence of polyploid races is usually accompanied by such phenomena, involving changes in geographical distribution.

The significance of these changes for the geographical distribution of plants is quite apparent. This is shown by the considerably greater ability of the new form to tolerate habitat conditions unusual for the initial form, such as low or high temperature, inadequate moisture, specific soil conditions, etc. Or, to put this in sequence of time: such conditions are the cause of the origin by mutation of a polyploid form adapted to the given extreme habitat conditions; here there arises the center of a new area, from which proceeds the dispersal of the new form over a territory frequently adjoining the area of the initial species but, due to its different ecological conditions, inaccessible to the latter.

Nevertheless, it would be erroneous to consider that only floras of regions characterized by extremes as regards ecological conditions embrace polyploid species and forms. There is no doubt that microclimatic conditions, micro-relief, and local edaphic conditions may also cause such mutant forms to appear. A study of the floras of the globe from this point of view should supply an answer to these questions. Work in this direction has only just begun, but, nevertheless, in the case of a number of species, there have already been published data of great interest for botanical geography.

If we take, for instance, the genus *Alopecurus*, comprising a polyploid series of species with chromosome numbers ranging from 14 to 108, we find that the alpine and arctic species have the largest chromosome numbers, which indicates that new polyploid species arose during the dispersal of this genus into mountainous and arctic regions (Strel-

KOVA, 1938). The same holds true for the genus *Agrostis*, which has a polyploid series of species with chromosome numbers ranging from 14 to 56 (SOKOLOVSKAYA, 1937). Or, if we take *Biscutella laevigata*, it has been shown that the subspecies that were established as initial on the basis of morphological data have a diploid chromosome number (18), while the derived species have a tetraploid chromosome number (36). The former occupy three distinct areas, situated, respectively, in the Rhine, Elba, and Danube Basins, while the tetraploid forms have a continuous area in the mountainous regions of southern Europe. The territory occupied by the latter area was covered with glaciers during the Würm period of glaciation, which indicates that the tetraploid forms arose in the post-glacial period as a result of the dispersal of the initial forms into mountainous regions after the recession of the glaciers. The diploid forms, which have been preserved outside the territory of maximum glaciation, are preglacial or interglacial relics (MANTON, 1934).

Both taxonomic and cytological data lead to the conclusion that the initial species of the section *Agrestis* of the genus *Veronica* are *V. filiformis* and *V. polita*, with 7 as their haploid chromosome number. Since the former species never left the confines of their initial habitat in the Caucasus, it may be presumed that the tetraploid species, *V. agrestis* and *V. opaca*, with n = 14, arose from *V. polita* after the latter's dispersal into northern Europe (BEATUS, 1936).

In the preceding examples the tetraploid species arose as a result of an extension of area of the initial species into colder regions. As an example of the reverse, *i.e.*, of the effect of high temperature, we may take the genus *Eragrostis*, which in the southern part of the Sahara Desert has an annual diploid species, *E. cambessediana* (n = 10), growing on humid, silt soils along the shores of small lakes, where the soil temperature amounts to not over 40° C. Near these same lakes, but outside the zone of inundation, there grows a species closely resembling *E. cambessediana* but perennial and tetraploid, *E. albida* (n = 20). Lastly, still further from the lakes on sand dunes, where the aridity of the soil and air attains its maximum and the soil temperature reaches as high as 80° C., *E. albida* also disappears, being replaced by a hexaploid (n = 40) species, *E. pallescens* (HAGERUP, 1931).

We wish, therefore, again to stress the fact that the utilization of cytological data for an understanding of the geographical distribution of plants, particularly for an elucidation of vicarious species and their areas, should constitute one of the important methods of historical plant geography. (For a more detailed discussion of this question *see* WULFF, 1937). However, this should not be understood in the sense that the origin of vicarious species is always connected with polyploidy. If we have discussed polyploidy in more detail, it has been only because the other modes of origin of vicarious species are either unknown or not yet well elucidated. In this connection the following words of DARWIN are still apropos: "No one ought to feel surprise at much remaining as yet unexplained on the origin of species, if we make due allowance for our profound ignorance on the mutal relations of the inhabitants of the world at the present time, and still more so during past ages" (*Origin of Species*, 6th ed., 1911, p. 156).

There is no doubt that vicarious species may arise not only as a result of autopolyploidy but also from various other causes. Thus, they may arise as a result of hybridization, likewise frequently accompanied by an increase in chromosome number (allopolyploidy), in which case the hybrids very often have areas that are not identical to the

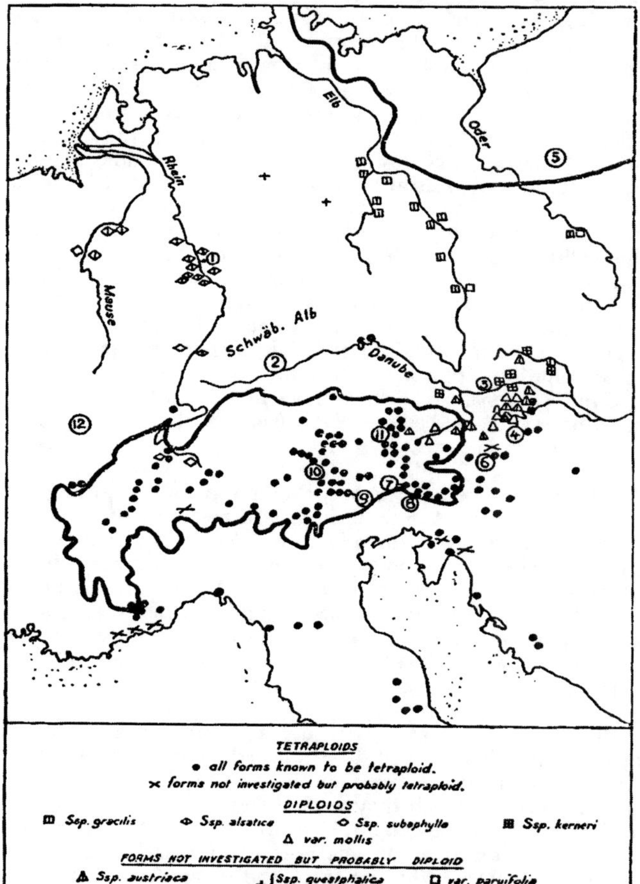

Fig. 9. — Distribution of diploid and tetraploid forms of *Biscutella laevigata*. (After Manton).

areas of the initial species, sometimes not coinciding in any part (*see* Gustafsson, 1935). Lastly, vicarious species may arise as a consequence of the differentiation of an initial form at one time widely distributed. Owing to changes in climatic and other habitat conditions in different parts of an originally continuous area, the initial form breaks up into a number of vicarious species, the origin of which may not be mutational, since during the process of natural selection intermediate stages may die out.

As the result of the breaking up of an initial type into a number of vicarious species, a geographical series of species is formed. In studying the area of any of these vicarious species for the purpose of understanding its origin, it is, therefore, necessary to investigate not only this area but also the areas of all the other species forming the series. As an example of such a geographical series let us take the series *Frutescens* in the genus *Caragana*, as given in KOMAROV's monograph (1908). The area of this series of species embraces all of China (except its tropical part), eastern and northern Tibet, Chinese and Soviet Turkestan, the western foothills of the Altais, West Siberia, and the southern and central part of European U.S.S.R. Within the limits of this area there is a series of vicarious species distinguished from one another both morphologically and ecologically: *Caragana chamlagu*, which grows in China proper, is replaced, as it approaches Mongolia, by the more xerophytic forms, *C. rosea*, *C. Leveillei*, and *C. opulens*. In Chinese Turkestan these are replaced by others still more xerophytic—the desert species, *C. polourensis* and *C. turfanensis*. In the region between Issyk-Kul and Zaisan there is a secondary center of species-formation, from which extend radially the areas of *C. lacta* and *C. Camilli Schneideri*, which have not spread beyond the boundaries of this region, and those of *C. frutex* and *C. grandiflora*, which have spread far to the west.

Relic Areas: — The relic concept in botanical geography may be formulated as follows: A relic species is a remnant of a more or less ancient flora having a relic area occupied by it at the moment of its entrance into the composition of a given present-day flora, from which moment the age of the relic is measured. A relic area is usually isolated and often contracting and discontinuous, constituting a remnant of a once-extensive area. These characteristics, however, are not always obligatory. A relic species—being, as DARWIN so aptly put it, a "living fossil"—is the embodiment of the historical development of a flora.

The difficult problem as to whether a certain species is a relic and to what period of time it should be referred may be solved by a study of the fossil remains of the given species and by a determination of the age of the deposits in which they were found, and, if such are not available, by indirect botanico-geographical methods. The latter should include a study of the areas of this relic species and of closely related species, not only within the limits of the given flora but throughout their entire extent, and also a general geographical analysis of the elements of the flora under study, which will elucidate the historical stages of development of this flora and the date of entrance into its composition of these or those elements.

In order, however, to elaborate a concept of relic species that will be generally accepted, it is necessary to come to an agreement regarding certain disputed points in their interpretation. These disputed points are as follows:

The term "relic" is commonly used in two different senses: taxonomic and botanico-geographical. In the former it refers to taxonomically isolated species belonging to ancient genera. Such species may also be geographically isolated, *i.e.*, they may have a relic area, but they may also be widely distributed at the present time and, consequently, not be relic in a geographical sense. A classical example of a taxonomically isolated species having a relic area is the maidenhair tree, *Gingko biloba*, while as an example of an ancient, taxonomically isolated species having at the present time an extensive area that is not relic we may take *Loiseleuria procumbens*. The latter is the sole species in the genus *Loiseleuria;* nevertheless, it has an arctic-alpine, for the most part post-glacial area. The use of the term "relic" in a botanico-geographical sense, on the other hand, always takes for granted a relic area, and consequently a relic species in this sense is always a remnant of an ancient flora, whether or not the genus to which it belongs is taxonomically isolated or not. Only in the latter, *i.e.*, geographical, sense should the term "relic" be employed, and then there would at once be eliminated various confusions resulting from its use. For remnants of ancient genera we might adopt the term proposed by SCHRÖTER (1934)—"Restanz". The establishment of the taxonomic antiquity of a species facilitates, of course, the establishment of its relic nature.

Differences of opinion as to the meaning of relic areas likewise arise from a different evaluation of the age of a relic. Here, also, the age of a relic should be understood not in a taxonomic but in a geographical sense. That is, in establishing the age of a relic species one should measure not from the date of its origin but from the date when it became part of the flora under study. Thus, if a species has formed part of a given flora since the Tertiary period, it will be a tertiary relic; if, however, it became part of this flora only during the Ice Age, it should be considered an Ice Age relic, irrespective of the fact that the species itself arose in the Tertiary period in another flora not under study. Consequently, the time a relic species has occupied its present habitat is the main criterion for establishing its age.

The date of the entrance of a species into the composition of any given flora may be determined only approximately, except when paleobotanic data provide definite proof of the existence of a species on a given territory uninterruptedly from the moment of the conversion of its remnants into a fossil state up to the present day. STOLLER (1921) maintains that only after such paleobotanic proofs of the uninterrupted habitation of a species may its relic character be accepted. Theoretically this is correct, but in practice it is rarely adhered to, since such proofs are possible only for an insignificant number of species. Consequently, it is necessary to resort to botanico-geographical methods and with their aid to draw conclusions as to the relic nature of a species.

If by the term "relic" we understand a remnant of a former flora, this does not mean that a relic species must necessarily occupy an area more limited at the present time than originally. The area of a relic may be comprised of a remnant or remnants of an extensive area no longer in existence, or, on the other hand, it may consist of an isolated, relic part of an extensive area, which itself may not be relic in character. In the latter case a species occupying the said area will be relic only in that part of its area which bears a relic character.

The main, non-relic part of an area may: (1) occupy its original territory, in which case the parts separated from it must at one time have constituted the periphery of the ancient area and have become isolated from the main part of the area due to changes in habitat conditions occurring within the limits of this peripheral region; or (2) be the result of a secondary dispersal of the species. In the first case the relic portions of the area constitute the remnants of a once-continuous but now greatly disrupted area. As an example of the second case we may take the common pine, *Pinus silvestris*, which in mountains has relic, detached portions of its area, while its extensive area on sandy soils is the result of a secondary dispersal.

It is usually considered that a relic area must be in a state of constant contraction, since a relic species is in disharmony with its present habitat conditions. Such a view of a relic area holds true only for certain relics and cannot be taken as a general rule for all. It holds good, for instance, in the case of *Cyclamen coum*, preserved only in one locality in the northern part of mountainous Crimea, where it finds itself in disharmony with existing climatic conditions, since it begins to flower only in November, when the Crimean winter is already well under way. Each year its flowers are destroyed by frost, and, consequently, its preservation in the Crimean flora to the present day is explicable only on the basis of vegetative propagation.

But there are many relic species which, as regards their biological peculiarities, have attained such a degree of adaptation to their present habitat conditions that their area is not now in the process of contraction and they are not becoming extinct. It is true that in most cases, if we regard the area of such a species not only in relation to that flora to which it now belongs but also in relation to its entire former area of distribution established on the basis of paleobotanic data, there can be no doubt that curtailment has occurred and also that the species has died out over all its former area with the exception of that part where it has been preserved and where, consequently, there did not take place changes in climatic or other habitat conditions in disharmony with the biological nature of the given species. For instance, the box tree, *Buxus sempervirens*, undoubtedly a Tertiary relic, has a much disrupted area, the various parts of which are characterized by dissimilar climatic and edaphic conditions. Nevertheless, this species grows in all parts of its area, which is not now in process of contraction; moreover, it does not occur isolated but as a co-member of a characteristic phytocoenosis, having become adapted to the specific habitat conditions in each separate part of its area (Christ, 1913).

However, the fact that the area of a relic species is not contracting is not always occasioned by its adaptation to new habitat conditions but may be due to the climatic conditions in the given area not having changed. Similarly, the isolation of an area may be the result of purely geomorphological causes or of changed climatic conditions in regions surrounding the given territory, while within the latter conditions have preserved their main features, if not completely, in considerable measure (*e.g.*, western Transcaucasia, as compared with other parts of the Caucasus adjacent thereto).

In many cases relic species do not extend beyond their limited habitats, since outside the boundaries they do not find favorable conditions for growth. For example, the relic species, *Wulfenia carinthiaca*, growing in isolated localities in the Carinthian Alps and on the Balkan Peninsula, does not extend beyond these localities. However, within the latter it is represented by millions of individuals, and it can by no means be considered a plant that is becoming extinct. It has been established that the confinement of this species to limited areas is due to its need of a humid climate (*see* GILLI, 1934, and literature cited by him).

Lastly, a relic species, which has inhabited a certain region uninterruptedly for a prolonged period, may have contracted and later again expanded its area, depending on climatic changes during geological ages. Thus, a Tertiary relic, inhabiting a certain territory from the Tertiary period to the present day, may have curtailed and again expanded its area corresponding to the great changes in climate—from Tertiary to glacial, from glacial to inter- and post-glacial—provided, of course, that this territory was located in that part of the globe where these changes affected habitat conditions.

The isolation of the stations of a species and the resultant discontinuity of its area are often regarded as definitely distinguishing the species as a relic. In many cases this holds true, but it is not an invariable rule, as a relic species may not have a discontinuous area and not every discontinuous area is relic, since the existence of isolated portions of an area may be due to biological and not historical causes.

A relic species may have an extensive area of distribution, embracing several natural regions of vegetation, or its area may be restricted to some one region, in which case the species may be regarded as an endemic relic. In case a species occupies an area relic throughout its entire extent, it may be called an absolute relic; if, however, only an isolated part of the area of a species is relic, the species is known as a local relic.

Not only separate species (single relics) or groups of species (relic groups or colonies) may be relic, but also entire floras. All the phytocoenoses composing the vegetation of a given region may be relic; on the other hand, there are relic phytocoenoses (associations, formations) belonging to floras that are not relic. Consequently, it is necessary to distinguish between *relic floras* (Refugialfloren or Primärfloren—SCHWARZ, 1938), most of the species composing which are relic, and floras destroyed as a result of climatic changes and restored due to the mass migration of species from surrounding floras—*migration floras* (Invasionsfloren—SCHWARZ, 1938).

A center of concentration of relic species is known as a *relic center*, which in some cases may coincide with the ancient center of development of the flora to which the given species belong.

If relic species find it possible to achieve a secondary distribution by the gradual occupation of habitats ecologically suited to them, such species may be called *migrant relics* (Wanderrelikte—SCHRÖTER, 1934). In such case, the more recently adopted habitats should be regarded, as first proposed by Scandinavian botanists, as pseudo-relic and the species themselves, in relation to the floras into which they have penetrated, as *pseudo-relics*. However, in their initial habitats these species remain true relics. Thus, V. I. KRECHETOVICH (1938) cites a number of species of the genus *Carex* having discontinuous areas, one part lying in northern latitudes and the other in the south in the mountains of the Caucasus and Central Asia. These species, in their southern habitats, he regards as pseudo-relics that became isolated on secondary post-glacial territories as a result of migration beyond the limits of the glaciers. This, however, is not the only meaning of a pseudo-relic; this term may also be applied in those cases when a present-day species acquires the apparent character of a relic.

Relic species under modern conditions are usually characterized by conservatism, by adaptation only to specific habitat conditions, in consequence of which they do not ordinarily expand their area or, if so, only to an inconsiderable extent.

A relic species, in case it finds itself in disharmony with present-day habitat conditions and occupies a contracting area, loses its capacity for variation and adaptation, and, hence, is in the process of becoming extinct. But, if it should chance upon favorable conditions, it may be restored to its normal state and give rise to new, polymorphic forms. Consequently, a relic species of economic value, upon being introduced by man into cultivation and transferred into an area potentially favorable, may become an economically valuable crop plant.

True relics, which have acquired their relic character as a result of natural causes, should not be confused with species that are becoming extinct due to the activities of man. Such species preserved from man's destruction are not true relics (unless, of course, they were relic prior to man's destructive activities); they may be called *anthropogenic relics* (cultivated relics, according to THELLUNG). Examples of such relics are the wild einkorn, *Triticum spontanum* Flaksb., and the wild emmer, *T. dicoccoides*. Ancient cultivated species, whose sown area, due to their low economic value, has been reduced to a minimum and which have been preserved only in a few localities, such as cultivated einkorn and emmer (*T. monococcum* and *T. dicoccum*), gourds (*Lagenaria vulgaris*), and others, may be called *cultivated relics*.

The chief factors in the origin of a relic area are, thus, changes occurring naturally, without the interference of man, in the habitat conditions within this area. On the basis of the fact that such changes may occur as a result of various causes, SCHRÖTER (1934) distinguishes three types of relics. The first he calls *formation relics;* these occupy limited areas within the boundaries of formations that have undergone considerable changes in their composition. For instance, the birch and *Juniperus foetidissima* within the limits of the beech district of the Crimea are both formation relics, but they differ in age, the juniper undoubtedly being a remnant of a formation occupying this territory during the Tertiary period, while the birch entered into the composition of the vegetation which replaced the Tertiary flora during the Ice Age. At the present time the birch grows in the Crimea only on a greatly restricted area to the number of a few hundred trees, whereas in former times it had a much wider distribution there. The latter fact has been definitely established by the finding, far to the north of the present distribution of the birch in the Crimea, of fossil remains of birch charcoal from the bonfires of prehistoric man. The birch was crowded out by the beech in the post-glacial period (Wulff, 1931).

The second type are the *geomorphological relics*, to which belong plants which are connected in their habitat with definite ecological conditions but which, due to geological and other historical causes, do not find themselves provided with the conditions of growth to which they are accustomed, *e.g.*, marine plants in fresh-water lakes, shore plants along the edges of former gulfs that are now dried up, species which have become isolated on islands formerly forming part of the mainland, etc. A very characteristic example of a geomorphological relic area is that of *Pinus eldarica*. This pine belongs to the group of Mediterranean species to which also belong *P. halepensis* and *P. brutia*, species characteristic of the Mediterranean region, and *P. pityusa*, growing in Asia Minor, the Caucasus, and the Crimea (ssp. *Stankevici*), to which last-mentioned *P. eldarica* is very close. These two species are, in turn, undoubtedly closely connected with the Tertiary pine, *P. sarmatica*, fossil remains of which have been found in Sarmatian deposits on the Kerch Peninsula. All these species grow for the most part along the seashore, with the exception of *P. eldarica*, which has a very limited area on slopes along the edge of the Eldar steppe in Transcaucasia. This steppe undoubtedly occupies the bed of a former sea, and the Eldar pines mark a shore line existing in an earlier geological epoch (Sosnovsky, 1928).

As another example of geomorphological relic areas we may take the isolated, islet-like habitats of *Quercus Ilex* on the southern slopes of the Alps near Lago di Garda, Trient, and Gemona, reaching to an altitude of 550 meters. The finding here, deep in the Alps, of this distinctly Mediterranean oak, ordinarily linked with marine climatic conditions, may be explained only by the relic character of these alpine habitats, preserved from a more extensive area existing in the Tertiary period, when the entire Po basin formed a gulf of the Adriatic Sea and the southern slopes of the Alps the shores of this gulf (Trotter, 1927). This type of relic may arise solely as a result of geomorphological changes unaccompanied by climatic changes.

Lastly, the third type of relics are *climatic relics*, *i.e.*, plants which, while growing now under certain climatic conditions, give evidence beyond any doubt that their origin and distribution took place under other climatic conditions. There are many such relics.

Due to the great diversity of types of relic species in nature, any rigid classification of them would, in our opinion, be artificial, and, hence, we do not consider it advisable to attempt to elaborate one. We wish merely here, in summing up, to emphasize that the basic factors determining the character of a relic species are its age, origin, and ecological type. The causes giving rise to the relic character of a species may be climatic, geomorphological, edaphic, or biotic. According to their age and origin, relic species may be subdivided into: (*1*) pre-Tertiary; (*2*) Tertiary (including (*a*) tropical and subtropical, (*b*) temperate, and (*c*) alpine-arctic); (*3*) glacial; (*4*) inter-glacial; and (*5*) post-glacial relics.

The species composing a flora may, thus, be divided into present-day elements and relic elements of various ages.

Continuous and Discontinuous Areas: — The area of a species, genus, or family may occupy a continuous territory or two or more separate territories. If the latter are so far separated that there is absolutely no possibility of their having been peopled by means of the dispersal of seeds by natural factors now existent, we call such areas *discontinuous*, in distinction from *continuous* areas occupying a continuous territory.

A continuous area may have ribbon-like prolongations (PACHOSKY, 1921) that may protrude beyond the main boundaries to a greater or less extent, following along rivers or mountain ranges that begin within the limits of the basic area and then traverse localities with entirely different habitat conditions.

According to HERZOG (1926), a discontinuous area is an area of any taxonomic unit that is broken up into several separate areas. ALPHONSE DE CANDOLLE defined the term "espèces disjointes" as "those species representatives of which, being found in two or more separate lands, nevertheless cannot be regarded as having been transported from one to the other because of some restraining circumstance—either the structure of the seeds, the mode of life of the plants, or the considerable distance between the lands inhabited" (1855, II, p. 993).

We have already seen that a species does not occupy every foot of territory of its range. Its topography may be very complex, giving grounds sometimes to regard the area as discontinuous, although in reality it is not. Such doubts may arise with respect to the areas of species adapted to a limited range of habitat conditions. For instance, species of fresh-water plants are distributed only where there are bodies of fresh water; a similar situation holds true as regards the vegetation of river valleys, swamps, mountain peaks, etc.

The "threshold of discontinuity" ("Disjunctions-Schwelle" in SCHRÖTER's terminology), *i.e.*, that distance beyond which any given taxonomic unit is unable to spread by natural means of dispersal, is very difficult to determine and cannot always be unconditionally established. Consequently, the discontinuity of an area considered to have arisen as a result of causes no longer extant may often be disputed, all the more since even views as to the significance and relative weight of factors at present in force and as to the capacity of plants for dispersal are widely at variance.

There are various types of discontinuity. Thus, if an area is composed of only two separate parts, one of which occupies an extensive territory and plays the dominant rôle, we may regard this as the *main part* and the other smaller part as the *subordinate part*, and the area itself as *bipartite* (BUSH, 1917). In other cases, when the area is broken up into a number of small, more or less equal parts, we speak of the area as *insular* and the discontinuity as *diffuse* (diffuse disjunction—SCHRÖTER). In such cases the discontinuity is manifested in the form of a much disrupted area, the cause for which is unclear but probably lies in the presence of very ancient types that at one time covered the area of distribution continuously but are now preserved only at separate points. Such a discontinuity is called *homogeneous* (BRIQUET), if in the separate parts of the area there grow identical forms. But very often in such separate parts of an area there grow not identical but related or vicarious forms; in such cases we propose the term *heterogeneous discontinuity*. The former of these two types of discontinuity is of more recent origin, the separated individuals not having yet had time to change their form, as they have in the second type, which is of more ancient origin.

To the type of heterogeneous discontinuity may be referred some cases of high-mountain discontinuity, which arose as a result of the formation on different mountain systems of vicarious species originating from one and the same widely distributed valley form. If this initial form should, owing to unfavorable habitat conditions, become extinct, these newly arisen vicarious forms would be entirely isolated. In such a case they are called *orophytes* (Oreophyten—DIELS).

There are other cases of high-mountain discontinuity, however, that are not heterogeneous but homogeneous, *viz.*, when there has occurred a simple breaking up of an area of a mountain species that at one time, possibly during the Ice Age, occupied a continuous area, and then with the change in climatic conditions died out over a considerable part of its range, being preserved only on isolated peaks and within the limits of several different mountain systems (*e.g.*, the area of distribution of species of *Euphrasia*).

In addition to high-mountain discontinuity where the parts of an area are located on different mountain peaks, there is *altitudinal discontinuity* where one part of an area is situated in one altitudinal zone and another part in another zone not directly adjoining the other, sometimes not even within the limits of the same mountain chain. Such discontinuous areas may arise as a result of the upheaval of mountain systems or the dispersal of a species followed by the development of high-mountain races. Usually the parts of an area that have been separated in such fashion are inhabited by vicarious species or races. For instance, the steppe species, *Scaligeria soongorica*, having spread northward from the southern part of Central Asia, is replaced—not in a steppe zone, where it is not found, but in a high-mountain zone—by a cold-resistant, sub-alpine species, *S. alpestris* (KOROVIN, 1934).

Another example of altitudinal discontinuity may be taken from a tropical flora. On the mountains of the Malay Archipelago, at an altitude of from 1,000 to 1,500 m., there grows *Bulbophyllum tenellum*, while another species very close to it, undoubtedly a vicarious species, *B. xylocarpi*, grows at sea level in mangrove woods (VAN STEENIS, 1935, p. 298—altitudinal vicarism).

An elucidation of the history of the origin of discontinuous areas is one of the main tasks of historical plant geography, since in this way there may be found the key to an understanding of many unclear moments in the history of floras. But, before we can undertake a study of their origin, we should become familiar with at least the main geographical types of discontinuous areas.

Chief Types of Discontinuous Areas:—

1. *Arctic-Alpine Type.*—This type of discontinuous area is characterized by species distributed, on the one hand, in the Alps and other mountains of central and southern Europe and, on the other, in arctic regions, in northern Europe and Asia, in the Altai Mts., and in North America. A few examples will illustrate the nature of this type:

Salix herbacea. Pyrenees, Alps, Apennines, Sudetes Mts., Carpathian Mts., Siebenbürgen, the Balkans—the Urals, northern and arctic Europe, mountains of England and Scotland, northern and arctic Siberia, northern and arctic North America.

Ranunculus pygmaeus. Central Alps, western Carpathians—northern Scandinavia, arctic Europe, Siberia, and America, and the Rocky Mts. as far south as 55° N.

Thalictrum alpinum. Pyrenees, Alps—Wales, Scotland, the Caucasus, northern and arctic Europe, Asia, and America, mountains of interior Asia, Rocky Mts. as far south as Colorado, Newfoundland, and Anticosti Island.

2. *North Atlantic Type.*—Plants distributed in North America and Europe (including the British Isles) and divided into two sections by the northern part of the Atlantic Ocean. Some of the species in this group are found also in isolated spots in Asia, including eastern Asia. Exceptionally interesting examples of this type are the areas of species found only in Ireland and North America, such as *Eriocaulon septangulare* and *Spiranthes romanzoffiana.* Other examples are:

Lycopodium inundatum. Northern and central Europe—North America.

Carex flava. Almost all of Europe, Anterior Asia—North America.

3. *Asturian Type.*—This type of area was noted as long ago as 1835 by WATSON and later by FORBES for such species as *Daboccia polifolia*, *Saxifraga geum*, and *S. umbrosa*, distributed in Ireland and then in southwestern France, the Pyrenees, Asturia, Cantabria, and Portugal (as regards *Dabeocia*, also on the Azores). A similar distribution has been established by SCHARF for a number of animals.

Such a discontinuity in the areas of these few species is all the more interesting since it resembles the discontinuity in many areas, where one portion embraces the British Isles and the other southwestern Europe and frequently also northern Africa (*e.g.*, the area of *Anagallis tenella*).

4. *North Pacific Type.*—This type of area is characterized by species now found, on the one hand, in Asia, particularly its eastern part (including Japan and Sakhalin), sometimes also in Europe, and, on the other, in North America, usually in the western (Pacific) section but in some cases in the eastern (Atlantic) section or in both sections. Discontinuous areas of this type are occupied by plants belonging either to one and the same species or to closely related species, the distinguishing characters of which arose as a result of prolonged isolation. In some cases the discontinuity in area has been established by the finding, in isolated regions, of species very close to those now living. As examples of this type we may take the areas of distribution of the following plants:

Liriodendron Tulipifera, Atlantic section of North America—*L. chinense*, China.

Hamamelis mollis and *H. japonica*, Japan—*H. virginiana*, United States east of the Mississippi (from the Gulf of Mexico to Canada).

Liquidambar orientalis, Asia Minor, and *L. formosana*, Formosa, southern China—*L. styraciflua*, Atlantic section of North America.

Catalpa bignonioides and *C. speciosa*, eastern U.S.A. from Illinois to Florida—*C. punctata* and *C. Kaempferi*, Japan, and *C. Bungei*, northern China.

Torreya nucifera, Japan, and *T. grandis*, China—*T. californica*, California, and *T. taxifolia*, Florida.

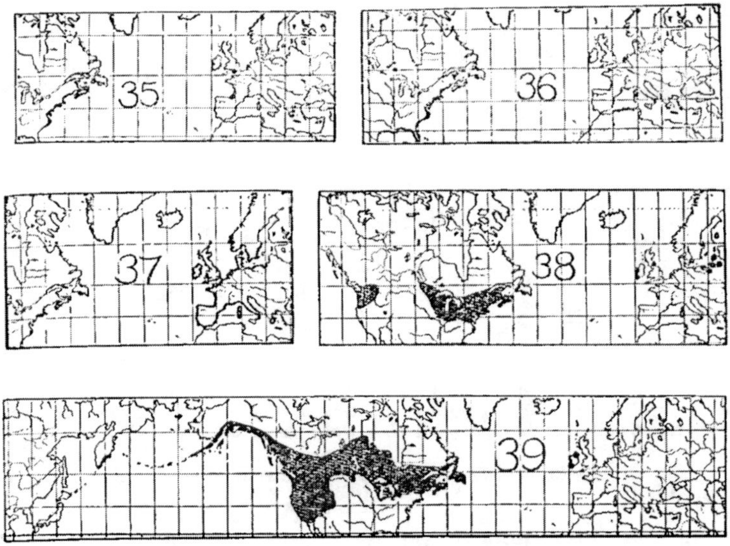

FIG. 10. — North Atlantic ranges of: 35, *Spartina alterniflora;* 36, *Spartina patens;* 37, *Puccinellia maritima;* 38, *Najas flexilis;* 39, *Spiranthes romanzoffiana.* (After FERNALD).

Osmorhiza brevistylis, Himalayas, northern Asia—North America from Canada to the mountains of Carolina in the East and to those of Mexico in the West; *O. laxa*, India, *O. amurensis*, Amur region, Kuznetz Alatau Mts., the Caucasus, and *O. japonica*, Japan—*O. nuda*, *O. brachypoda*, and *O. longistylis*, Atlantic section of North America, and *O. Berterii*, *O. chilensis*, and *O. depauperata*, South America (Chile, Argentina). All the species of this genus are close to one another and form a semicircle, extending from the Himalayas over the Bering Sea and along the Cordilleras to southern Chile.

Sequoia, a genus represented now by but two species, *S. gigantea* and *S. sempervirens*, growing only in California. In a fossil state representatives of this genus have been found throughout all of North and part of South America, and also in Europe, Spitsbergen, and Asia.

Taxodium, represented now by three species, *T. distichum*, *T. imbricarium*, and *T. mexicanum*, distributed along the south Atlantic coast of North America and in Mexico. In a fossil state it is known throughout almost all of North America and also in a considerable part of Europe and Asia.

Glyptostrobus, a genus very close to *Taxodium* and represented now by but two species, *G. pendulus* and *G. heterophyllus*, found in eastern Asia (China). In contrast to the two preceding genera that died out in Asia, *Glyptostrobus* died out in America. In a fossil state (from the Cretaceous) it is known in Greenland—*G. groenlandicus* and *G. intermedius*. From Tertiary deposits two species are known: *G. europaeus*—from southern Europe to the arctic zone and in North America;

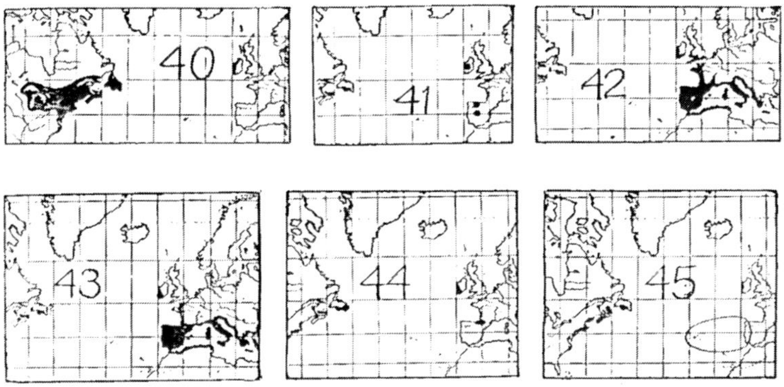

FIG. 11.—North Atlantic and Iberian ranges of: 40, *Eriocaulon septangulare;* 41, *Erica Mackaii;* 42, *Rubia peregrina;* 43, *Arbutus unedo;* 44, *Saxifraga geum;* 45, the genus *Corema*, *C. alba* (European), *C. Conradii* (American). (After FERNALD).

G. Ungeri—Switzerland, Spitzbergen, Siberia, North America, and Greenland.

5. *North America-South America Type.*—This type is characterized by identical or closely related species found both in North and South America but not having a continuous area. Here belong also those species and genera that are represented in one of the Americas by living forms and in the other by extinct forms. As an example of the latter we may take the genus *Sequoia*, growing now in California but in a fossil state found also in Chile. Among plants now growing in both North and South America but not having continuous areas we may take as an example the *Sarraceniaceae*, which family includes the closely related genera *Sarracenia*, *Darlingtonia*, and *Heliamphora*. The genus *Sarracenia* has six species distributed in the Atlantic section of North America; *Darlingtonia* one species, *D. californica*, in California (Sierra Nevada Mts.); and *Heliamphora* one species, *H. nutans*, in British Guiana.

6. *Europe-Asia Type.*—This type of area is characterized by identical or related species found both in Europe and Asia but not

having continuous areas, despite the fact that these two continents are united. There are very many examples of this type, of which we shall cite but a few:

Cimicifuga foetida. Central Europe, southwestern part of European U.S.S.R.—southern Siberia from Novosibirsk Region to the Pacific Ocean, China, Japan, the Himalayas.

Leontice altaica. Southern part of European U.S.S.R.—eastern part of Semipalatinsk Region, southwestern part of Novosibirsk Region.

Middendorfia borysthenica. Southern part of European U.S.S.R., Kirghiz steppes (Airtau), Kara-Irtysh Valley—Novosibirsk Region (one habitat).

Cymbaria borysthenica. Crimea, Dnepropetrovsk and Nikolayev

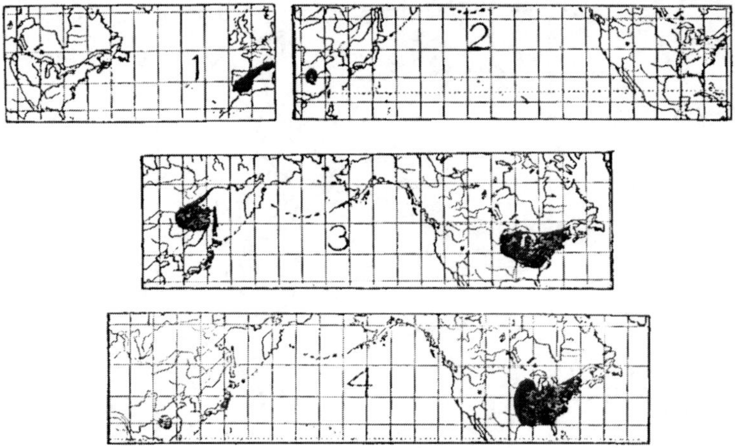

FIG. 12. — North Atlantic and Pacific ranges: 1, *Vallisneria spiralis* (European) and *V. americana* (American); 2, *Liriodendron Tulipifera* (American) and *L. chinense* (Chinese); 3, *Symplocarpus foetidus;* 4, *Polygonum virginianum.* (After FERNALD).

Regions in the Donetz Valley—*C. dahurica*, Siberia westward as far as the Yenisey, Mongolia to the Chinese border, Manchuria, and *C. mongolica*, southwestern Mongolia (perhaps this species should be combined with the preceding).

Wulfenia carinthiaca, Carinthian Alps, Balkan Peninsula (mountains of Albania and Montenegro), and *W. Baldacci*, Albania—*W. orientalis*, Syria, and *W. Amherstiana*, western Himalayas, Afghanistan.

7. *Mediterranean Type.*—This type comprises many diverse variations, seven of the most important being: (*a*) discontinuous areas in Mediterranean lands territorially connected; (*b*) discontinuous areas of which one portion lies along the northern coast of the Mediterranean Sea and the other in northern Africa; (*c*) one portion located on the mainland of the Mediterranean Basin and the other on islands in the Mediterranean Sea; (*d*) one portion on the mainland of the Mediter-

ranean Basin and the other on islands in the Atlantic Ocean (Azores, Madeira, Canary Is.); (*e*) one portion in southern Italy and the other in the Balkans; (*f*) one portion in the Balkans and the other in Asia Minor; (*g*) one portion in the Crimea, a second in Transcaucasia, and a third in Asia Minor. Among the many other variations of this type we may mention discontinuous areas having one portion in the Mediterranean Basin and the other in the mountains of tropical Africa, or in Cape Province, or in California, etc.

8. *Tropical Types.*—In the tropics of both the Old and New Worlds we find, just as in the preceding vegetation zones, a number of types of discontinuous areas characteristic of the plants in this flora. The existence of such discontinuities is particularly evident in areas of

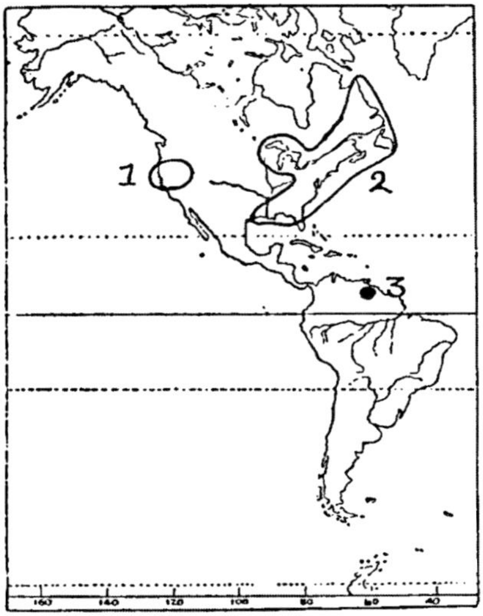

Fig. 13. — North-South American discontinuous area of *Sarraceniaceae;* 1, *Darlingtonia;* 2, *Sarracenia;* 3, *Heliamphora.* (After Hutchinson).

so-called pan-tropical species, *i.e.*, those now distributed in tropical and subtropical regions of Asia, Africa, and America. But, in addition, we have within the tropics a number of more localized discontinuities, which we may designate as *paleotropical discontinuities*, occurring in the areas of species and genera distributed in tropical Africa and Asia, and also in some cases in Madagascar, Polynesia, and the Malay Archipelago and *neotropical discontinuities*, occurring within the limits of the tropics of the New World. Among the more important tropical types of discontinuous areas we list the following:

a. *Asia-Africa Type.*—To this type we refer the areas of plants common only to the floras of Africa and continental Asia. It is a characteristic fact that in western Africa we find a large number of species of tropical India; in eastern Africa, on the other hand, we find the predominance of more widely distributed species, resembling Mediterranean forms. As examples of this type we may cite the areas of the following genera: *Ancistrocladus, Sesamum, Citrullus, Podocarpus, Pandanus, Borassus, Musa, Olea, Coffea, Luffa, Lagenaria, Momordica, Anisophyllea, Hypericum* (*H. mycorens*).

b. *Africa-Madagascar Type.*—In the case of this and the following type we understand under "Madagascar" not only Madagascar proper but also the adjacent group of islands (Reunion, Mauritius, Seychelles). As an example of this type of area we may cite *Viola abyssinica*, found in Abyssinia and the Cameroons and then on Madagascar and adjacent

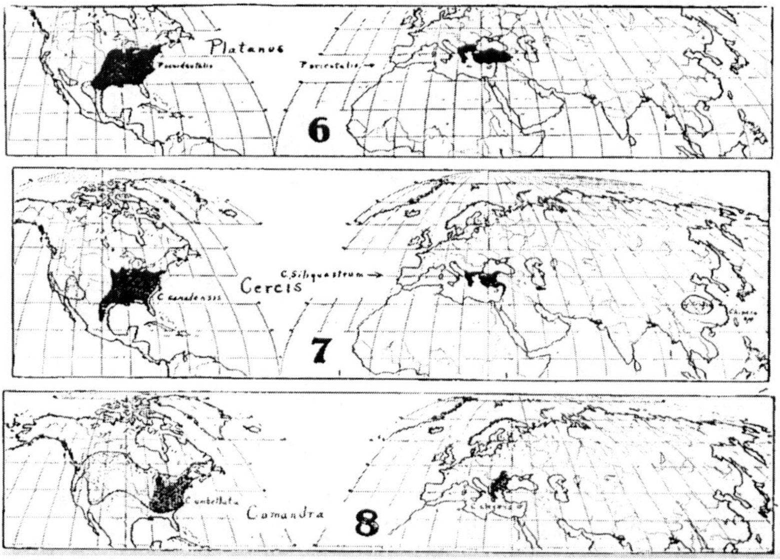

FIG. 14. — Mediterranean-North American discontinuous areas: 6, *Platanus orientalis* (Mediterranean), *P. occidentalis* (North American); 7, *Cercis Siliquastrum* (Mediterranean), *C. canadensis* (North American); 8, *Comandra elegans* (European) and *Comandra umbellata* (North American). (After FERNALD).

islands (here represented by a vicarious or perhaps even identical species, *V. emirnensis*). *Geranium simense, Senecio Bojeri*, and others have similar areas.

c. *Asia-Madagascar Type.*—Among genera distributed both on Madagascar and in Asia (India), extending sometimes also to Australia and Polynesia, we may mention *Nepenthes, Wormia*, and *Pothos*.

In addition, there are a number of species and genera growing in Madagascar that are found also both in Africa and India, *e.g.*, *Buddleia madagascariensis*.

d. *Africa-America Type.*—The tropics of Africa and America are characterized by a number of genera found only on these two continents and absent in Asia, or more widely distributed on them than in tropical Asia. As examples of type of area we may cite the following:

The genus *Symphonia* of the family *Guttiferae* is represented by the species *S. globulifera*, a small tree growing in western Africa from Gabon to Angola, and then is distributed in the forests of Brazil and also in the Guianas, Panama, Costa Rica, Honduras, and the West Indies. This species is not found in eastern Africa, but in Madagascar there grow five other closely related species.

As a second example we may take the family *Vochysiaceae*, which is purely American except for the genus *Erismadelphus*, recently found in the forests of the Cameroons.

The genus *Anona*, most of whose species are distributed in America but which is likewise cultivated in Africa, is also represented on the latter continent by wild species, such as *A. glauca* in upper Guinea, *A.*

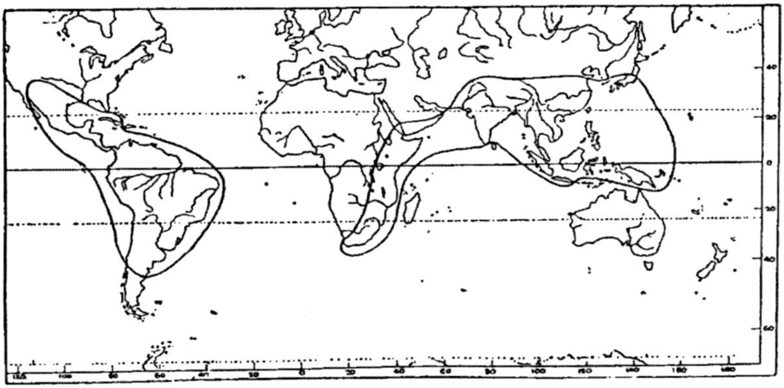

FIG. 15. — Pantropic discontinuous area of the genus *Buddleia* (*Loganiaceae*). (After HUTCHINSON).

Klainii in Gabon, and *A. senegalensis*, closely related to the American species, a wild form found in all the steppe regions of Africa.

e. *India-Malay Type.*—This type embraces the discontinuous areas of species distributed in India, Malaysia, Polynesia, and also in some cases in tropical Australia, particularly the northeastern part. As examples of this type we may cite the genera *Agathis*, *Araucaria*, *Dacrydium*, *Areca*, *Engelhardtia*, *Cochlospermum*, *Hernandia*, and *Cratoxylon*.

Here we may refer also the discontinuous areas of plants found in Malaysia, Polynesia, Australia, and New Zealand. According to OLIVER, in the flora of the last-mentioned country we find a considerable number of elements that are common to both New Zealand and Malaysia. For instance, the genus *Dacrydium*, represented by several species in Malaysia and New Caledonia, has 7 species in New Zealand and 1 in Tasmania; it also has 1 species in Chile, *D. Fonckii*, close to the New Zealand species, *D. laxifolium*.

9. *Gondwana Type.*—In this group we include, in conformity with paleogeographical data, those discontinuous areas embracing India,

Africa, Madagascar, and Australia. There are not many examples of this type in the plant kingdom, particularly among the *Angiospermae;* nevertheless, there are a few characteristic areas, chiefly for Africa, Madagascar, and Australia. As examples we may cite *Adansonia, Keraudrenia, Rulingia,* and *Athrixia.*

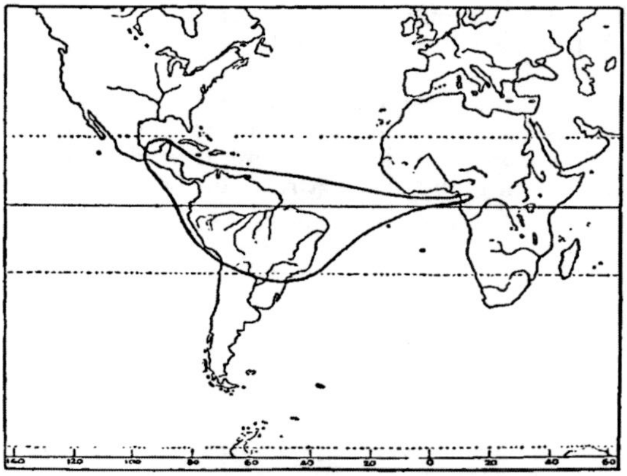

FIG. 16. — Neotropical discontinuous area of the *Vochysiaceae* (in Africa one genus, *Erismadelphus*). (After HUTCHINSON).

10. *South Pacific Type.*—Here are included the areas of genera and species common to South America, Pacific islands, New Zealand, and Australia, or found in New Zealand and not in Australia. Among such areas are: *Jovellana:* South America—New Zealand; *Pernettya, Hebe, Nothofagus:* South America—Australia, New Zealand; *Drimys:* South America—Australia and Polynesia.

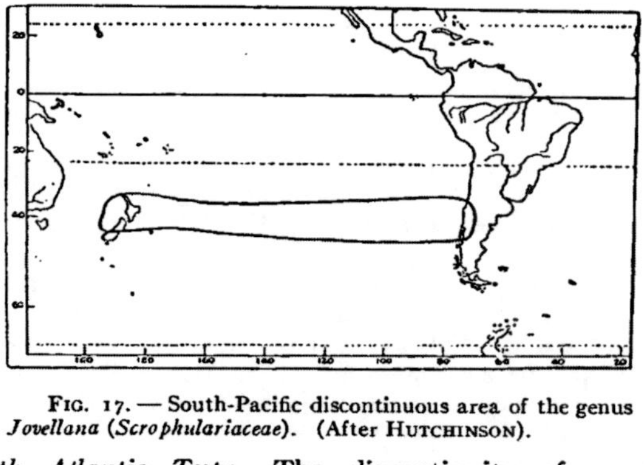

FIG. 17. — South-Pacific discontinuous area of the genus *Jovellana* (*Scrophulariaceae*). (After HUTCHINSON).

11. *South Atlantic Type.*—The discontinuity of areas of many genera found in South America and also in Africa and Madagascar was

noted long ago. According to ENGLER, such discontinuous areas of genera found only in South America, Africa, and Madagascar may be divided into three main groups. To the first he refers genera having numerous species distributed in South America, of which only a few are found also in Africa, and monotypic genera distributed both in South America and Africa, related genera of which are found in South America. For instance, *Telanthera* (*Amaranthaceae*) has 45 species in South America, of which only one is found in western Africa; *Paullinia* (*Sapindaceae*) has 80 species in South America, only one of which grows in Africa.

To the second group are referred genera having maximum distribution in South America, but a few species not found in South America grow in Africa. For example, *Copaifera* (*Leguminosae*) has 10 species in South America and 2 in Africa, of which one is related to

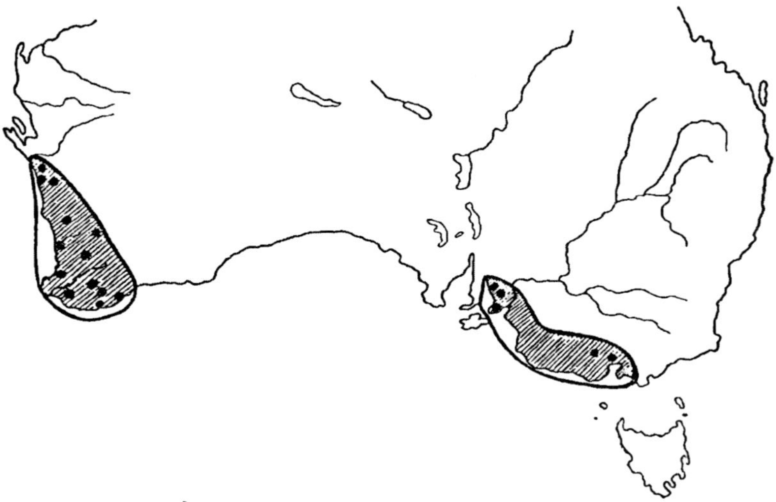

FIG. 18. — Australian discontinuous area of the section *Erythrorhiza* of the genus *Drosera*. (After DIELS).

a species found in Cuba; *Asclepias* (*Asclepiadaceae*) has 58 species in South America and 2 in Africa. Here we may refer also the very interesting area of the genus *Ravenala*, distributed on Madagascar and in South America.

To the third group are referred genera having several species in South America and several in Africa or represented in their floras by a single species each. For instance *Sphaeralcea* (*Malvaceae*) has 21 species in South America and 4 in Africa (Cape Province); *Amanoa* (*Euphorbiaceae*)—2 species in South America and also 2 in western Africa; *Chlorophora* (*Moraceae*)—1 species in tropical America and 1 in Africa.

12. *Australian Type.*—Within the limits of Australia there are numerous and very interesting cases of discontinuous areas, having separate portions in eastern and western Australia, the intervening zone in central Australia forming a botanico-geographical region known as Eremea. This zone varies greatly in size and shape, as a result of which the discontinuous areas have diverse configurations.

13. *Antarctic Type.*—Here are included the discontinuous areas of species and genera found on the Antarctic mainland and also in the southern part of South America, New Zealand, Australia, South Africa, and on the various islands scattered between them—St. George, Kerguelen Land, Macquarie Island, etc. It is understood that the parts of these areas located on the Antarctic mainland are now represented by plants found only in a fossil state. As examples of this type of area we may mention the areas of the genera *Nothofagus* and *Fitzroya.*

The above-enumerated thirteen types do not by any means cover the entire diversity of discontinuous areas. It suffices to mention the existence of areas having one portion in the northern and one in the southern hemisphere. For instance, in Tierra del Fuego and southern Patagonia there are found a number of boreal plants; also in Australia (*e.g.*, the genus *Veronica*); and so on. We have not attempted to list all the types of discontinuous areas, since that would be well nigh impossible, but we believe that the types enumerated show quite clearly their great diversity and the need of similarly diverse explanations as regards their origin.*

References:

BEATUS R., 1936: Die *Veronica*-Gruppe *Agrestis* des Sektions *Absinebe.* Ein Beitrag zum Problem der Artbildung (Zeitschr. ind. Abst. und Vererb., Vol. 71, No. 3).

BRAUN-BLANQUET, J., 1923: L'origine et le développement des flores dans le massif central de France (Paris-Zürich).

BUSH, N. A., 1917: The most important terms employed in floristic phytogeography (In Russian; Jour. Russ. Bot. Soc., Vol. 2, No. 1-2).

CHEVALIER, A. and GUÉNOT, L., 1925: Biogéographie (*In* MARTONNE, "Traité de géographie physique", Paris).

CHRIST, H., 1913: Über das Vorkommen des Buchsbaumes (*Buxus sempervirens*) in der Schweiz und weiterhin durch Europa und Vorderasien (Verhandl. Naturf. Ges. Basel, Vol. 24).

CLEMENTS, F. E., 1934: The relict method in dynamic ecology (Jour. Ecol., Vol. 22, No. 1).

DE CANDOLLE, ALPH., 1855: Géographie Botanique Raisonnée, Vols. I, II (Geneva).

DIELS, L., 1906: Die Pflanzenwelt von West Australien (*In* ENGLER u. DRUDE, "Veget. d. Erde", Vol. VII, Leipzig).

DIELS, L., 1908: Geographie der Pflanzen (Berlin; 3rd ed., 1929).

---

* IRMSCHER (1929) makes an attempt to classify discontinuous areas, and we give below the more important types included in this classification (for further details *see* IRMSCHER). The figure 1 stands for North and South America, 2 for Europe-Africa, 3 for Asia, and 4 for Australia.

1) Areas distributed in 1.2
2) " " " 1.2.3.
3) " " " 1.3.
4) " " " 1.3.4.
5) Southern types of discontinuity:
   a) 1.4.
   b) 1.2.
   c) 2.4.
   d) 1.2.4.
6) Areas distributed in 1.2.3.4.
7) " " " 2.3.4.
8) " " " 2.3.
9) " " " 3.4.
10) Areas limited to some one continent (1, 2, 3, or 4).
11) Areas of plants distributed on oceanic islands.

DRUDE, O., 1890: Handbuch der Pflanzengeographie (Stuttgart).
ENGLER, A., 1879, 1882: Versuch einer Entwicklungsgeschichte der Pflanzenwelt, Vols. I and II (Leipzig).
FURSAEV, A. D., 1937: On the problem of species-formation along the banks of rivers (In Russian; Soviet Bot., No. 3).
GILLI, A., 1934: Die Ursachen des Reliktcharakters von *Wulfenia carinthiaca* (Engl. Bot. Jahrb., Vol. 66).
GOOD, R. D'O., 1927: Summary of discontinuous generic distribution in the Angiosperms (New Phytologist, Vol. 26, No. 4).
GUSTAFSSON, A., 1935: The importance of the apomicts for plant geography (Bot. Notiser, Vol. 5).
HAGERUP, O., 1932: Über Polyploidie in Beziehung zu Klima, Oekologie und Phylogenie (Hereditas, Vol. 16).
HAGERUP, O., 1933: Studies on polyploid ecotypes in *Vaccinium uliginosum* L. (Hereditas, Vol. 18).
HERZOG, TH., 1926: Geographie der Moose (Jena).
ILYINSKY, A., 1933: Area and its dynamics (In Russian; Soviet Bot., No. 5).
IRMSCHER, E., 1922, 1929: Pflanzenverbreitung und Entwicklung der Kontinente, I and II (Mitt. Inst. Allg. Bot., Hamburg, Vols. 5, 8).
KASHKAROV, D. D. and KOROVIN, E. P., 1931: An analysis of the economic routes of migration of the flora and fauna of Soviet Central Asia (In Russian; Jour. Ecol. and Biocoen., Vol. I).
KERNER, A., 1869: Die Abhängigkeit der Pflanzengestalt von Klima und Boden, ein Beitrag zur Lehre von der Entstehung und Verbreitung der Arten; gestüzt auf die Verwandtschaftsverhältnisse, geographische Verbreitung und Geschichte der *Cytisus*-Arten aus dem Stamme *Tubocytisus* DC. (Innsbruck).
KOMAROV, V. L., 1901: Flora Manshuriae, Vol. I, Chap V. The Species and its Subdivisions (In Russian; Acta Horti Petropolitani, Vol. 20).
KOMAROV, V. L., 1908: Introduction to the floras of China and Mongolia (In Russian; Bull. Bot. Gard. St. Petersb., Vol. 29).
KOROVIN, E. P., 1934: Types of migration in the plant world (In Russian; Bull. Cent. As. State Univ., Ser. 8, No. 16).
KRECHETOVICH, V. I., 1938: On the problem of the glacial pseudo-relic *Cyperaceae* in the floras of the Caucasus and Central Asia (*In* "The Relic Problem in the Flora of the U.S.S.R.", Part I; in Russian; Acad. Sci. U.S.S.R., Moscow-Leningrad).
LAVRENKO, E. M., 1936: On the problem of the age of psammo-endemism in the southern part of European U.S.S.R. (In Russian; Bull. State Geog. Soc., Vol. 68, No. 1).
LEHMANN, E., 1937: Die Gattung *Veronica* in entwicklungsgeschichtlicher Betrachtung (Cytologia, Fushii Jubil. Vol. 2).
LEWITSKY, G. A. and KUZMINA, N. E., 1927: The karyological method in the taxonomy and phylogeny of the genus *Festuca* (subgenus *Eu-Festuca*) (In Russian; Bull. Appl. Bot., Gen. and Plantbr., Vol. 17, No. 3).
MANTON, I., 1934: The Problem of *Biscutella laevigata* L. (Zeitschr. ind. Abst. u. Vererb., Vol. 67, No. 1).
MOLISCH, H., 1926: Pflanzenbiologie in Japan, Kap. XIV. Über den Kosmopolitismus der Pflanzen (Jena).
PACHOSKY, I., 1921: Fundamentals of Phytosociology (In Russian; Kherson).
Relic Problem in the Flora of the U.S.S.R. 1938, Parts I and II (In Russian; Acad. Sci. U.S.S.R., Moscow-Leningrad).
RIDLEY, H. N., 1923: The distribution of plants (Ann. Bot., Vol. 37).
RIDLEY, H. N., 1925: Endemic plants (Jour. Bot., Vol. 43).
SCHRÖTER, C., 1912: Genetische Pflanzengeographie (Handwörterbuch d. Naturwiss., Vol. IV, Jena; 2nd ed., Jena, 1934).
SCHWARZ, O., 1938: Neue Ergebnisse der Phytochorologie (Chronica Botanica, Vol. 4, No. 1).
SOKOLOVSKAYA, A. P., 1937: Karyo-geographical investigation of the genus *Agrostis* L. (In Russian; Bot. Jour., Vol. 22, No. 5).
SOSNOVSKY, D., 1928: Present status and future outlook of the State Forest Reserve of *Pinus eldarica* (In Russian; Bull. Azerbaidzhan Univ., Vol. 7).
STEENIS, C. G. G. J. VAN, 1935: On the origin of the Malaysian mountain flora (Part 2, Bull. Jard. Bot. Buitenz., Ser. III, Vol. 13, No. 3).
STOLLER, J., 1921: Die Pflanzenwelt des Quartärs (*In* POTONIÉ–GOTHAN, Lehrbuch der Paläobotanik, 2nd ed.).
STRELKOVA, O., 1938: Polyploidy and geographo-systematic groups in the genus *Alopecurus* L. (Cytologia, Vol. 8, No. 3–4).
TROTTER, A., 1927: Un relitti di flora mediterranea nell'alto Fruili occidentale (Atti R. Ist. Veneto di Sci., Lett. ed Arti, Vol. 86, No. 2; cited in Engler's Bot. Jahrb., 1930, Vol. 63, No. 2).
VIERHAPPER, F., 1919: Über echten und falschen Vikarismus (Oest. bot. Zeitschrift, No. 1).

WAGNER, M., 1868: Die Darwin'sche Theorie und das Migrationsgesetz der Organismen (Leipzig).

WAGNER, M., 1870: Über den Einfluss der geographischen Isolierung und Kolonienbildung auf die morphologische Veränderung der Organismen (München).

WAGNER, M., 1880: Über die Entstehung der Arten durch Absonderung (Kosmos, Nos. 1, 2, 3).

WANGERIN, W., 1924: Beiträge zur Frage der pflanzengeographischen Relikte (Abh. d. Naturf. Gesellsch. in Danzig, Vol. I, No. 1).

WETTSTEIN, R., 1898: Grundzüge der geographisch-morphologischen Methode der Pflanzensystematik (Jena).

WILLIS, J. C., 1936: Some further studies in endemism (Proc. Linn. Soc. London, Session 148, Part 2).

WULFF, E. V., 1926: The origin of the Crimean flora (In Russian; Bull. Crimean Soc. Naturalists, Vol. 9).

WULFF, E. V., 1931: Die Birke in der Krim (Mitt. Deutsch. Dendr. Ges.).

Chapter VI

PARALLELISM IN THE GEOGRAPHICAL DISTRIBUTION OF PLANTS AND ANIMALS AND CORRELATION BETWEEN THE DISTRIBUTION OF PARASITES AND THAT OF THEIR PLANT HOSTS

The difficulties involved in retracing the history of the geographical distribution of plants, considering the paucity of paleobotanic data, make it necessary, in addition to the direct study of the areas of species, to make use of all possible indirect methods that may aid in solving this problem. Among such methods the chief is a comparison with data on the geographical distribution of animals.

The distribution of plants and animals, both at the present time and in former geological epochs, is in many ways interdependent. Ecological conditions, past changes in climate and in the connections and configurations of continents, are reflected, to a greater or less extent, in the distribution of all living organisms. Consequently, the geography of plants and animals really should constitute a single science, biogeography. Such a science is now only a dream for the future, but it is to be hoped that it will some day be realized. If now the study of the geography of the floras and faunas of the globe is separated, this is the regrettable result of the scope of modern knowledge, which has outgrown the possibility of its mastery by a single investigator. This need for specialization hampers to some extent the study of nature as a unified whole.

It is, nevertheless, of utmost importance that phytogeographers and zoogeographers should co-operate closely in their work. The agreement of data in both branches of biogeography, the similarities in the distribution of plants and animals, the occurrence of the same discontinuities in their areas, and the identity of the centers of origin of floras and faunas leave no doubt as to the existence of historical causes for these phenomena and give a sound basis for phyto- and zoo-geographical conclusions. In the present chapter we do not aim to go into this phase of our general problem in detail. We wish merely to give a few illustrations of parallelisms in the distribution of plants and animals, which will make clear the importance of checking conclusions made on the basis of a direct study of plant distribution by comparison with zoogeographical data.

As a second indirect method, we wish to stress the importance of a study of parasites—both of plant and animal origin, connected during the course of their life cycles with definite plants—which may provide very valuable aid in establishing the history of the distribution of their plant hosts. The parallel study of both reveals their interrelations, paths of migration, and also their past areas of distribution, often indicating that there formerly existed a different distribution of the dry lands and climatic zones of the globe.

By the aid of such indirect methods it frequently becomes possible to establish the past distribution of plants in cases where ordinary methods—data on present distribution and paleobotany—fail to provide sufficient clues.

Parallelism in the Distribution of Plants and Animals: — Similar features in the distribution of plants and animals arise from three circumstances: (*1*) the dependence of animals on plants as a source of food and of plants on animals as agents in the transfer of pollen and the dispersal of fruits and seeds; (*2*) the effect of identical ecological factors and the habitation of common biocenters (this constitutes the basis for the view held by some investigators that there do not exist separate plant and animal associations but only common, complex bio-associations); and (*3*) historical causes, such as changes in climate and in the position of the continents, having a like effect on the areas of plants and animals.

In connection with the theme of our present manual of particular interest are those similarities in distribution induced by historical causes. These are expressed in the frequent occurrence of relic retreats common for both plants and animals. Moreover, sometimes a plant association, *e.g.*, a forest, may be a relic retreat for animals, although it itself is not relic in nature. This is confirmed by the fact that a number of plants and animals have the same centers of development of their areas and also by the similarity of their discontinuous areas. This may be seen from the following examples. According to GEPTNER (1936), twenty-one species of fresh-water fish of the British Isles, for which the sea constitutes an insurmountable barrier, are identical with fish living in the Rhine and rivers of northwestern France. The occurrence of the same plants in Great Britain and along the banks of the Rhine is a parallel phenomenon, and is explained by the fact that the Rhine, at the time of the post-glacial connection between the British Isles and the mainland, flowed through England and had its mouth in the northern part of Norfolk County.

On a plateau in the central part of the Sahara there is found fauna of the Mediterranean type, similar to the fauna of the Atlas Mts., despite the fact that deserts now separate them. Similar discontinuous areas are characteristic of the floras of the mountain chains of Africa. The absence of marked changes in climate since the beginning of the Tertiary period in the region of the Indian Ocean has resulted in the presence on its islands of an exceedingly rich flora, preserving Tertiary features, and also of a very rich fauna, including a shore fauna that arose from the fauna of Tethys. There are animals that have bipolar areas with a break in the tropical zone, just as in the case of plants.

COCKERELL (1932) has pointed out an interesting parallelism between the discontinuous area of a genus of bees, *Hesperapis*—distributed in arid districts of southwestern U.S.A. and also, under similar habitat conditions, in South Africa—and an analogous area of a genus of plants, *Menodora*, of the family *Oleaceae*. This genus is distributed in semi-arid districts of the southwestern part of North America (including Mexico and the southwestern States of the U.S.A.), in central and southern South America, and then, after another break, in the

Transvaal and other adjacent parts of South Africa. Out of seventeen species and thirteen varieties of *Menodora*, STEYERMARK (1932), a monographer of this genus, found that "nine species and eleven varieties occur in North America, six species and one variety in South America, and two species and one variety are found in Africa, one of the species, *M. heterophylla*, being found in North America and having its variety in Africa. There are, accordingly, three distinct areas of distribution: (*1*) southwestern United States and Mexico, (*2*) central and southern South America, and (*3*) South Africa" (p. 100). All these regions are characterized by arid or semi-arid conditions. STEYERMARK considers that the area of the genus *Menodora*, and consequently of the bees connected therewith, "in all probability had a more continuous geographical range at least before the end of the Cretaceous period", and that "the most logical source of evidence explaining this present interrupted distribution lies in the postulation of a land-bridge once connecting South America and Africa" (p. 104).

A number of identical discontinuous areas of animals and plants are cited by REINIG (1937), who summarizes the data of Dr. L. S. BERG and others in his book, "Die Holarktis". Among such areas are those having the following distribution: Europe-eastern Asia, Eurasia-America, and Europe-North America. REINIG believes that these areas had their origin as a result of conditions during the Ice Age.

Evidence for the existence of a land-bridge between Africa and India is provided not only by floristic but also by faunistic data. Thus, SCOTT (1933), on the basis of a comparison of the insect fauna and flora of the Seychelles and adjacent islands, comes to the conclusion that these islands "are not typical oceanic islands, but of an ancient continental type. . . . The high percentage of endemic forms in both flora and fauna indicates a biological association which has endured since the remote past, but which has existed throughout the archipelago as a whole; there has been little formation of distinct species in the individual islands, the separation of which has probably been comparatively recent. . . . However many be the endemic plants and animals which have died out, those which persist are the remains of the flora and fauna of a much larger, almost vanished, land" (p. 381). The presence of a considerable percentage of species of plants and animals having related forms in Africa, on the one hand, and India, on the other, indicates that this archipelago constitutes the remnants of a land-bridge formerly connecting Africa and India. The fact that "the element with African affinities in the endemic fauna is of more recent origin than the element with Indo-Australian affinities" (p. 382) points to the longer persistence of a connection with Africa, that with India having been broken at a considerably earlier period.

The foregoing examples suffice to show what great significance for historical plant geography may be the checking of floristic data with those of zoogeography.

Rust Fungi as an Index of the Distribution of their Plant Hosts: — In the works of MORDVILKO (1925, 1926) we have valuable data obtained by a comparison of the distribution of rust fungi and of their hosts. In this case the possibility of botanico-geographical conclusions is based on the fact that the spring generation of rust fungi form on their plant hosts so-called aecia. These aecia develop only under specific conditions existing on their usual host but not on any other plants. Moreover, this specificity of rust fungi is so great that it sometimes enables us to distinguish between forms or even species of plants in which with the naked eye we are unable to detect any difference in morphological structure. The summer and fall generations of these fungi are, however, much less specialized, and, consequently, the uredospores and teliospores formed by them may develop on various other plants. There thus may come about a facultative heteroecism of the rust fungi that may later become permanent. If the new hosts prove to be a more favorable substratum, the summer generations usually develop on them, and thus an originally autoecious parasite becomes heteroecious, although often, besides heteroecious forms, the initial autoecious forms are preserved. Hence, in cases where there are only summer and fall generations of a rust fungus on secondary hosts, we may assume that they arose from autoecious forms in places where the initial host grows, on which passed the entire cycle of development of the fungus, including the spring aeciostage.

According to MORDVILKO, the cycles of rust fungi evolved in this way: At first there were "lepto-forms" with several generations of teliospores, among which might be distinguished the summer forms and the forms that wintered over. Then came "brachy-forms", when the summer teliospores were replaced by uredospores; and, lastly, there arose the "eu-forms", when the first generation of uredospores is replaced by aecia. Only "eu-forms" could develop heteroecism. This occurred in the following manner: When in a given region there arose new plants, or plants from other regions migrated to this region, the rust fungi passed from their former hosts to new ones, provided the latter offered them favorable conditions for development, and in time they became modified into new species and then genera. But such a passing over to new hosts could be achieved only in the case of "lepto-forms" and "brachy-forms". As regards "eu-forms", it would be more complicated, since only the generations of uredospores and teliospores could adapt themselves to the new hosts, while the aecia could develop only on the primary hosts. At first, consequently, there arose non-obligatory or facultative heteroecism and only later regular, obligatory heteroecism. Hence, it would follow that the primary hosts are more ancient than the secondary, and in most cases this has proved to be so.

If the foregoing is kept in mind, then, by the aid of the history of rust fungi, it becomes possible also to elucidate certain problems in the history of their hosts—to decide which are of a more ancient and which of a more recent origin, which are indigenous to a given region and which adventive. Cycles could evolve only in a temperate climate with its seasons, while for the tropics with their even climate only "lepto-forms" with a single type of spore are characteristic, all other forms of rust fungi (with uredospores or aeciospores) undoubtedly being immigrants from regions with a temperate climate. A study of tropical rust fungi from this point of view might also give much of value for an elucidation of the history of their hosts.

The possibility of the utilization of the described biological peculiarities of rust fungi for purposes of historical plant geography may be illustrated by the following examples:

Microcyclic forms of rust fungi may arise from heteroecious forms, when in any country or region—as a result of a change in climate (*e.g.*, from temperate to cold or vice versa, from humid to arid or vice versa) —their primary hosts, on which the aecia develop, die out, but their secondary hosts, on which the generations of uredospores and teliospores develop, survive. In such cases the fungi multiply by means of the uredospores (these generations can also winter over); the teliospores, however, at first are preserved in a rudimentary state (due to the dying out of the primary hosts, they lose all their significance), and then completely disappear. In this way there arise microcyclic forms that cannot convert themselves into macrocyclic forms, even if in the given locality their primary hosts should again make their appearance. In western Europe, for instance, the rust fungus *Cronartium quercium* propagates itself on oaks only by their uredospores, the teliospores having completely disappeared, but in North America and Japan the same fungus passes through a complete cycle, the aecia developing, moreover, on two-leaved pines. It seems that in Tertiary times there existed in Europe at least two species of pine resembling the North American species on which aecia of *Cronartium quercium* are formed—*Pinus rigios* (Oligocene-Miocene), similar to *P. Taeda*, and *P. hepios* (Oligocene-Pliocene), similar to *P. mitis*. Or let us take *Thecospora vacciniorum*. This rust fungus propagates itself on *Vaccinium Myrtillus* and *V. Vitis-idaea* only by uredospores (teliospores are rare), but in North America the same fungus passes through a complete cycle, the aecia being formed on *Tsuga canadensis*. But in the Pliocene period in Europe there existed *Tsuga europaea*, which might well have been the primary host of *Thecospora vacciniorum*. In the European part of the U.S.S.R., in connection with the disappearance of the fir (*Abies*) and the larch (*Larix*) that had existed here in interglacial times, there arose a considerable number of microcyclic *Melampsoraceae*.

Sometimes macrocyclic and microcyclic forms are found side by side, *e.g.*, of *Puccinia graminis* in Great Britain. The microcyclic form (only uredospòres on cereals) could have arisen only in the following manner: The host, the barberry (*Berberis vulgaris*), which existed there in pre-glacial times, disappeared during the Ice Age, and there arose a microcyclic form (minus teliospores). Later, however, when in Great Britain the barberry again made its appearance (believed by some to have been introduced by man), there reappeared with it the macrocyclic form of *Puccinia graminis*. These examples show that microcyclic forms of fungi may give us a clue as to which plants (the primary hosts of the fungi) died out in a given region due to climatic changes that occurred there in former times.

Very interesting data are presented by LASHCHEVSKAYA (1927), who investigated the rust fungus that occurs on the Tertiary relic, *Schivereckia podolica*, growing in a part of Kursk Region characterized by a number of such relics. It was found that this rust fungus is *Puccinia drabae*, a species of an exclusively mountain or north arctic-alpine type. The hosts of this fungus are ordinarily several species of the genus

Draba, that grow in the mountains of central Europe, in the Alps, throughout the entire Arctic region, in North America, in the Caucasus Mts., in the northern part of European U.S.S.R., in the Urals, in western Siberia (the Altais), in Kamchatka, in the mountains of Central Asia, and in the Himalayas. *Schivercckia* is a genus very closely related to *Draba* and distributed in the Urals, in Siberia, and on the Podolian elevation. The finding of this rust in the Kursk Region indicates that in the past—presumably in the Ice Age—there existed there conditions analogous to those prevailing in its present habitats. This, in turn, confirms the relic, not adventive, character of the distribution of *Schivereckia podolica*.

Rust fungi, having presumably evolved simultaneously with their hosts, have, in case the latter are characterized by discontinuous areas, analogous discontinuous areas, being found on the same or related species of the host plants. V. A. Transhel (1936, 1940) cites a number of very interesting instances of such parallel discontinuity of areas, a few of which we shall give here. Thus, the rust genera *Uredinopsis*, *Milesina*, and *Hyalopsora* pass their life cycle on different hosts, forming aecia on the fir (*Abies*) and uredospores and teliospores on species of genera of ferns belonging to the families *Polypodiaceae* and *Osmundaceae*. It is of interest that these genera of rust are found chiefly in North America and eastern Asia, thus repeating the discontinuous area characteristic of the higher plants that are their hosts.

Species of *Puccinia*, not identical but related, occurring on the genus *Lycium* and the related genus *Grabowskia*, that grows in Palestine, South Africa, and North and South America, are characterized by the peculiar structure of their teliospores, the pedicels of which swell greatly in water. These species of rust presumably arose and became adapted to the mentioned hosts prior to the appearance of the discontinuities in the areas of these genera. Subsequently, with the differentiation of the hosts there also occurred differentiation of the species of the parasite. One of the latter, *Puccinia Thuemeniana*, is parasitic on *Myricaria germanica* in the Tyrol and Rumania and on *M. dahurica*, a vicarious species, in Siberia, in the Altai and Sayan Mts. Similarly, *Phragmipedium circumvallatum* is found on species of *Geum* of the sections *Orthurus* and *Oligocarpa* in Spain, the Caucasus, and Central Asia.

These examples suffice to show how a study of fungi may be used to establish the history of the development of their hosts. This method of investigation, however, has as yet been developed but little and is rarely used.

Plant-Lice as an Index of the Distribution of their Hosts: — Just as fungi so also insects parasitic on plants may in precisely the same way give valuable clues to the history of the distribution of their plant hosts. To Mordvilko (1928, 1929, 1930) credit is likewise due for elucidating the interesting biogeographical interrelations between plant-lice and the plants upon which they pass their life cycles.

In plant-lice, just as in rust fungi, there are a number of different cycles of alternating generations, in this case a succession of virgin and bisexual generations. The evolution of plant-lice proceeded in the direction of the replacement of winged bisexual forms first by winged virgin forms and later by wingless virgin forms. Of insects of the bisexual generation first the females become wingless and then the males.

Present-day plant-lice represent different stages in the evolution of cycles of plant-lice: some are only at the beginning of this evolution; others at its end. Such cycles develop only in temperate climes with their different seasons. In the tropics, with their even climate, the bisexual generation is lacking, and there remain only winged and wingless asexual forms. Whenever during the history of the earth there arose new groups of plants or into a given region there penetrated new plants from some other region, plant-lice have migrated from their old hosts to new ones, if the latter were suitable for them, *e.g.*, from willows and poplars to maples, from birches to maples, from *Myricaceae* and *Juglandaceae* to *Tilia*, to various *Papilionaceae*, etc., and here on the new hosts they became transformed first into new species and then into new genera.

Such a migration from certain plant hosts to others leads eventually to heteroecism. On the primary host there develop only some of the generations, including the bisexual generations and those asexual generations developing from fertilized eggs that have wintered over (so-called "fundatrices"), and on the new, secondary host there develop the summer asexual generations that terminate in the formation of winged sexual forms (sexuparae), which fly back to the primary host to deposit their eggs.

Consequently, in plant-lice, just as in rust fungi, the dying out of their primary hosts in any region leads to the formation of microcyclic forms, represented only by wingless and winged asexual generations. For instance, in the case of the disappearance of *Ulmus* and *Pistacia*, that serve as primary hosts of plant-lice, the latter are preserved on the roots of cereals or other plants in the form of asexual generations. Hence, on the basis of the distribution of microcyclic forms we may judge as to the former distribution of their primary hosts. For example, the species of plant-louse *Forda formicaria* has as its secondary host the roots of various cereals. A microcyclic form is now distributed throughout the entire holarctic (Europe, Siberia, North America); presumably in Tertiary times *Pistacia terebinthus* also had approximately the same area of distribution.

Another plant-louse, *Trifidaphis phaseoli*, has as its primary host *Pistacia mutica* and as its secondary host the roots of various dioecious herbs. As a microcyclic form it is distributed from Central Asia and Egypt through the southern part of eastern Europe and through western Europe to Great Britain, Greenland, North America, and even South America (it is not found in Siberia). This gives grounds for assuming that in Tertiary times *Pistacia mutica*, or a form very close to it, had approximately the same area of distribution. Pistacia lice (microcyclic forms) are found also in Formosa, Java, Egypt, and South Africa. Moreover, the finding of the microcyclic form of *Trifidaphis phaseoli* on the shores of Greenland, taking into account the fact that this plant-louse spreads very slowly, indicates, without doubt, that it has existed in Greenland uninterruptedly from the time that there grew there either *Pistacia mutica* or some closely related species of this genus. This louse, therefore, must have lived through the Ice Age in Greenland, which gives grounds for presuming that not all representatives of the *Angiospermae* were destroyed by the great climatic changes of that period. The secondary hosts on which this louse passed its life cycles must also have been preserved in sheltered places not covered by the ice. That this louse penetrated into Greenland from Europe, by way of Iceland, is not possible, since it is not found in Iceland. In that country there is found another louse, *Pemphigus bursarius*, whose primary host is the poplar, *Populus nigra*. It spreads considerably more rapidly than the preceding species; nevertheless, it is not found in Greenland (MORDVILKO, 1935). These facts fully coincide with recent data on the origin of the flora of higher plants in Greenland, pointing to the likelihood that some of its species lived through the Ice Age.

As a final example, we may take the plant-louse, *Forda hirsuta* (primary host—*Pistacia vera*). The fact that the microcyclic form of this louse is not found in Europe shows that *Pistacia vera* never occurred on the territory of Europe.

The data we have here presented make evident the need for more detailed studies of the interrelations between plant and animal organisms from the point of view of their geographical distribution.

References:

COCKERELL, T. D., 1932: Discontinuous distribution in plants (Nature, Vol. 130, No. 3291).

GEPTNER, V. G., 1936: Principles of Zoogeography (In Russian; Moscow-Leningrad).

LASHCHEVSKAYA, V. I., 1927: A contribution to the problem of the origin of the flora of the Kursk-Orel plateau: *Puccinia drabae* Rud. and *Schivereckia podolica* (Boss.) Andrz. (In Russian; Bull. Sci. Res. Inst., Voronezh Univ.).

MORDVILKO, A., 1925: Analozyklische Uredinales und ihr Ursprung (Biol. Centralbl., Vol. 45, No. 4, pp. 217–231).

MORDVILKO, A., 1925/26: Die Evolution der Zyklen und die Heterözie bei den Rostpilzen (Centralbl. f. Bakteriol., Parasitenkunde u. Infektionskrankheiten, Vol. 66, pp. 505–531).

MORDVILKO, A., 1928: The evolution of cycles and the origin of heteroecy (migrations) in plant-lice (Annals and Magazine of Nat. History, Ser. 10, Vol. 2, pp. 570–582).

MORDVILKO, A., 1929: Analocyclic elm aphids, *Eriosomea*, and the distribution of elms during the tertiary and glacial periods (Comptes Rendus Acad. Sci. U.R.S.S.).

MORDVILKO, A., 1929: Die analozyklischen Pistazien-Blattläuse und die Verbreitung der Pistazien in der Tertiärzeit (Comptes Rendus Acad. Sci. U.R.S.S.).

MORDVILKO, A., 1930: *Pemphigus bursarius* Tullgren (*pyriformis* Licht.) and its analocyclic forms (Comptes Rendus Acad. Sci. U.R.S.S., pp. 50–54); Aphids of the Subtribe *Hormaphidina* (*Ibid.*, pp. 168–172).

MORDVILKO, A., 1930: On the origin of heteroecy in plant-lice (Comptes Rendus Acad. Sci. U.R.S.S., pp. 256–260).

MORDVILKO, A., 1935: Die Blattläuse mit unvollständigem Generationszyklus und ihre Entstehung (Ergebn. u. Fortschr. d. Zoologie, Vol. 8).

REINIG, W. F., 1937: Die Holarktis. Ein Beitrag zur diluvialen und alluvialen Geschichte der zirkumpolaren Faunen u. Florengebiete (Jena).

SCOTT, HUGH, 1933: General conclusions regarding the insect fauna of the Seychelles and adjacent islands (Trans. Linn. Soc., Ser. 2, Zoology, Vol. 19, Part 3).

STEYERMARK, G. A., 1932: A revision of the genus *Menodora* (Ann. Mo. Bot. Gard., Vol. 19, No. 1).

TRANSHEL, V. A., 1936: Rust fungi as indicators of the relationship of their hosts in connection with the phylogenetic development of rust fungi (In Russian).

TRANSHEL, V. A., 1940: The Rust Fungi of the U.S.S.R. (In Russian; Moscow-Leningrad).

Chapter VII

ARTIFICIAL FACTORS IN THE GEOGRAPHICAL DISTRIBUTION OF PLANTS

In examining the factors responsible for plant distribution, it is necessary, first of all, to establish whether those aspects of their distribution not explicable on the basis of present-day physiographic conditions—isolated fragments of areas, relic and endemic species and centers, various combinations of floras, etc.—are to be explained as effects of that most potent of modern factors, the activity of man, or of more ancient, historical factors. If the former be the case, we have no need to delve into complicated paleogeographical and paleoclimatological problems; if the latter, we should concentrate our attention primarily on such problems and on other natural factors.

When changes in the distribution of plants are induced, accidentally or intentionally, by the action of man, we say that they are caused by "artificial factors". Instances of man's influence on vegetation are so numerous and well known to every one that there is no need of recounting them in detail. We shall here consider only general problems connected with man's influence on plant distribution, with the aim of determining the significance of man's activity as an explanation of those phenomena in the geographical distribution of plants which are of particular interest for our branch of science.

Man's activities not only have an indirect effect on plant distribution, *e.g.*, by the destruction of natural plant communities, but man acts as a direct agent in the transfer of plants from one locality to another. Moreover, even as a direct agent, we may draw a distinction between man's intentional change of habitat of a plant, often by transfer over very great distances, and his involuntary aid to a plant's change of habitat. In the former case we have in mind plants introduced by man into cultivation—cultivated plants and their derivatives; in the latter—weeds and other wild plants that have accompanied man during his migrations over the globe.

With respect to cultivated plants, we shall here take under consideration only cases of their naturalization and subsequent reversion to conditions of an independent existence, to a so-called "wild" state, becoming what are known as "escapes". The reversion of cultivated plants to a wild state may take place in various ways. In some cases the plants continue to grow only in the same place where they were formerly cultivated. Such plants are known as *anthropogenous relics*. In other cases they may, as a result of various natural factors, become distributed beyond the limits of the place where they were once cultivated. This has from ancient times not infrequently taken place, particularly in localities where the natural vegetation has previously been destroyed and habitat conditions disturbed by man. However, the number of such plants, for reasons which we shall take up later, is not very large.

Man has played a considerably greater rôle in the distribution of weeds, which have either accompanied the crops grown by man or directly accompanied man in his migrations. These are plants possessing in most cases a wide range of tolerance, so hardy and unfastidious that they can spread rapidly far and wide.

All lands now have their specific weeds, a study of the origin of which reveals a remarkably close connection with the trade routes and migrations of man. Thus, North America has at present about 500 species of weeds brought from Europe. All the American representatives of a number of genera—*Lamium*, *Malva*, *Medicago*, *Melilotus*, etc.—had their origin in Europe and bear an adventive character.

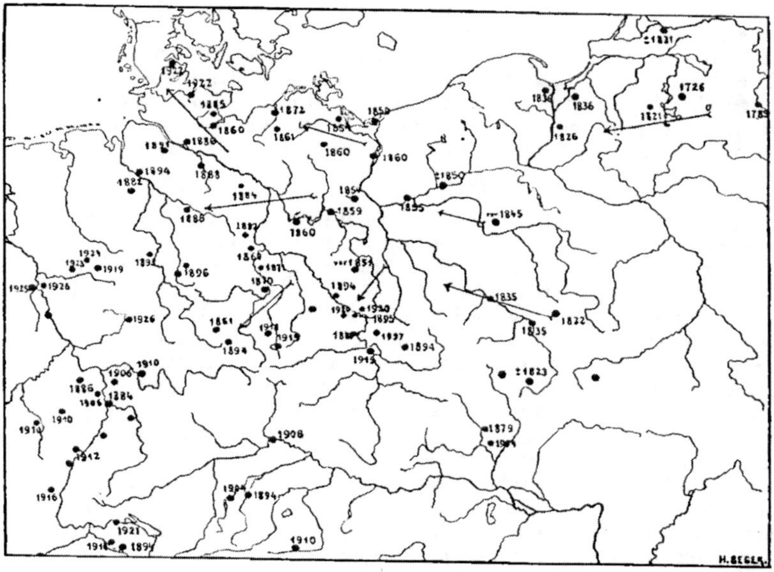

FIG. 19. — Map showing the gradual migration of the adventive plant, *Senecio vernalis*, into western Europe from the East beginning with the year 1726. (After BEGER, from HEGI).

Conversely, Europe has a number of species that were brought from America, such as *Erigeron canadensis*, now common throughout almost all of Europe, and *Elodea canadensis*, which has become so widely distributed in European waters that it constitutes an obstacle to shipping.

The large number of weeds in existence was, until recently, taken as proof of the predominant significance of man as a factor in the geographical distribution of plants. Basing themselves on the many, diverse possibilities for the transfer of plants by man and the great abundance of weeds in existence, the extreme advocates of the rôle of man concluded that all aberrations in plant distribution—discontinuous areas, isolated habitats, relics out of harmony with present-day vegetation and habitat conditions—may be explained as a direct or indirect effect of man's activities. And, as in the case of any such extremism, the most fantastic assumptions and unfounded statements were regarded as scientific and well-founded. The bald statement that the distribution of a given plant is the result of its introduction by man was considered sufficient to refute all propositions regarding the effect of historical causes on the present distribution of plants.

What limits the significance of man's activities as a factor in the geographical distribution of plants? First, the short duration, as compared to the antiquity of the plant world, of the action of man on the latter, and, second, the fact that, even conceding, as we do, the great rôle of man in the transfer of seeds and the spread of weeds, this does not determine the extent to which this vegetation is able, in places where it penetrates into regions of the earth to which it is not native, to gain a foothold in the local flora and become a full-fledged member. The few instances of the almost cosmopolitan distribution by man of some plants do not vouchsafe the unlimited possibility of the naturalization of plants, regardless of what habitat conditions are normal for them and what geographical obstacles may be in the way of their further dispersal.

Loss by cultivated plants (and by weeds infesting them) of their ability to maintain an independent existence and to spread independently: — First of all, there arises the question as to whether a plant, once in cultivation, can pass, after it is no longer cultivated, into a wild state, *i.e.*, can maintain an existence independent of man. In other words, does not man's influence on a plant under cultivation result in so changing its structure that it is, to a certain extent, deprived of the advantages in the struggle for existence that it had acquired during the entire preceding period of its life history? As regards those cultivated plants utilized by man for centuries and modified by breeding and selection to conform to man's needs, it is quite apparent that the answer to this question must be affirmative. The changes wrought by man in the structure of plants may be exceedingly diverse. We shall here note only the most important, taking as our chief sources the works of ZINGER (1909), HILDEBRAND (1873), and THELLUNG (1915).

1. *Conversion of perennials into annuals*, as an effect of cultivation, is not an uncommon phenomenon. We know, for instance, among cereals many cases of cultivated annuals that have originated from wild, perennial forms. Thus, the wild progenitors of cultivated rye were perennials (BATALIN, 1892), and the wild rye, *Secale montanum*, closely related to cultivated rye, is also a perennial.

2. *Increase in the size of seeds*, accompanied by a decrease in their number and, consequently, also in the number of embryos, greatly lessens the plant's chances for propagation.

3. *Loss by fruits and seeds of their protective coverings*. For example, naked cereals have fruits deprived of the outer glumes that served to protect them. The lining of sclerenchymatous tissue in the pods of wild *Leguminosae* is lacking in cultivated species.

4. *Loss of natural adaptations for the dissemination of seeds*. Here

we may cite the many-jointed spikes of cereals (so-called "brittle" spikes), a character common among wild species but absent or nearly so in cultivated species, and the bipartite fruits of wild species of *Umbelliferae*, a character only very slightly expressed in cultivated species. Plants which in a wild state have fruits that, when ripe, dehisce in one way or another, thus giving the wind a chance to disperse the seeds, may lose this ability in a cultivated state. For instance, the fruits of cultivated flax or of the opium poppy do not dehisce when ripe, while those of the wild poppy, *Papaver Rhoeas*, and of the wild flax, *Linum crepitans*, are dehiscent.

The loss of the sclerenchymatous tissue in the pods of *Leguminosae*, about which we have already spoken, lessens the extent to which the valves of the pod curl up, thus diminishing the ability of the plant to scatter its seeds.

In general, the lesser development of sclerenchymatous tissue in many crop plants is a common phenomenon, apparently connected with the fact that plants, when cultivated, are grown in closer stands, are often protected from winds, and are in other ways provided with special local habitat conditions characterized by greater humidity and more shading, all of which induce more vigorous growth (ZINGER, 1909). Thus, plants of a more or less xerophytic nature and growing under xerophytic climatic conditions may, by cultivation, be transformed into mesophytes, but these latter, when left uncultivated, become entirely unviable. A similar phenomenon we observe in trees. In a forest the trees have tall, slender trunks and weakly developed crowns, while, when grown farther apart, they have short, sturdy trunks and vigorous crowns. The latter easily resist winds, whereas forest trees, when isolated, *e.g.*, by the cutting out of trees, are often blown down by heavy winds.

5. *Double flowers*, induced by the conversion of stamens into petals, constitute a clear example of an aberration rendering the affected plants incapable of life without man's assistance, except, in certain cases, through adequate vegetative means of propagation.

6. *Loss of various defensive adaptations*, such as thorns, spines, pubescence, etc., making the plant defenceless against herbivorous animals and more susceptible to injury from evaporation of moisture.

7. *Exceptionally great increase in size of fruits*, markedly distinguishing cultivated from wild plants, may not be of advantage to the plant, since it appears to render the fruits more difficult for animals to eat and, consequently, constitutes a hindrance to seed dissemination.

8. *Improvement in the flavor of fruits* tends to cause animals to eat them up completely, seeds and all.

9. *Seedless (parthenocarpic) fruits*, just as double flowers, render the independent existence of a plant absolutely impossible.

All the above-mentioned peculiarities, occurring also among wild plants as abnormalities, are, by means of conscious or unconscious selection, intensified and highly developed in the case of cultivated plants.

10. *Complete absence of fruiting* in some cultivated plants—due to atrophy of the sex organs (as, for instance, to a considerable degree, in the sugar cane), to absence of pollinators in places of cultivation far distant from the plant's native habitat (*e.g.*, *Yucca* species in Europe),

or to other causes—results in these plants becoming incapable of completing their sexual cycles.

When no longer cultivated, all plants such as we have enumerated above—modified by man in ways most suited to his needs but abnormal and harmful for the plant itself as an independent organism—naturally tend to disappear and are no longer able to aid in extending the area of the species to which they belong.

Of exceptional importance is the circumstance that the very same changes in biological peculiarities and morphological structure that we have just described in plants purposely cultivated may also be observed in those weeds that constantly accompany definite crops, thanks to which they are involuntarily cultivated by man. Certain of such weeds, without man himself being aware of the process, may gradually become transformed into direct objects of his cultivation. This has undoubtedly been the mode of origin of cultivated rye, oats, hemp, and coriander, and most probably of several *Leguminosae*, kenaf (*Hibiscus cannabinus*), Chinese jute (*Abutilon Avicennae*), etc.

As proof of the existence of such involuntary breeding of weeds, making possible their subsequent cultivation, we may cite some of the numerous instances of the occurrence in weeds, the planned selection of which by man is entirely excluded, of changes in structure and biology analogous to those we listed above for cultivated plants. THELLUNG (1915) compiled considerable data of this sort. Here we learn of the involuntary selection of weeds for annual habit, due to the annual plowing of the soil, and of the consequent transformation of perennial plants into annuals, an example being *Phalaris brachystachys*, a Mediterranean weed differing only in being annual from its close relative, *P. truncata*, that grows in natural habitats. In weeds, just as in cultivated plants, we may observe an increase in the size of seeds at the expense of their number, and the loss by fruits of their protective coverings and of their ability to disseminate their seeds naturally.

Examples of such changes may also be taken from ZINGER's work (1909) on species of *Camelina* infesting flax. *C. glabrata* and *C. linicola* constantly accompany flax, being known as "*plantae linicolae*". According to ZINGER, these plants "do not have a single one of the adaptations, possessed by most wild annuals, for accommodating their development to seasonal changes. They are unable to await the moment most favorable for the development of stem and flowers, having neither rosettes like their nearest wild relatives nor seeds that may lie long dormant like many other annual plants. Their fruits ripen almost simultaneously, instead of successively over a prolonged period, which would assure the preservation of at least that part of the progeny attaining by the onset of winter a stage of development favorable for wintering-over. These species can normally complete the cycle of their development only provided their seeds are gathered when ripe, kept in a storehouse during the winter, and sown in the spring."

The well-known, widely distributed weed, *Agrostemma githago*, has many-seeded pods, which, although dehiscent, are so constructed that the seeds cannot fall out by themselves. According to BRAUN-BLANQUET, the spikes of *Bromus secalinus*, a weed infesting cereals, and its close relative, *B. grossus*, break up into sections only late in the season and then not completely, while the very closely related species, *B. racemosus* and *B. pratensis*, have such brittle spikes that they break up into separate spikelets very readily. Moreover, the awns in *B. secalinus* are much reduced in length or rudimentary, which is undoubtedly connected with the loss of ability and need for independent distribution. *Alectrolophus apterus*, a weed infesting rye, has lost the ability of self-dissemination due to the fact that its seeds do not have any wing-like appendages. Its seeds ripen simultaneously with the rye seeds, and they are, thus, harvested and sown together with the latter.

Loss of those adaptations making independent existence possible leads in some cases to the plant ceasing altogether to grow in a wild state and to its becoming so closely linked with some crop plant that it is as difficult to determine the wild progenitors of such a cultivated weed as it is in the case of many crop plants. As examples of such weeds of unknown ancestry we may cite two infesting cultivated flax — *Silene linicola* and *Cuscuta epilinum.*

The above facts serve to illustrate one of the limitations to the significance, from a botanico-geographical standpoint, of man's part in the distribution of plants. But this is not the only limitation.

Acclimatization. — Of no less importance, as regards the possibility of plants becoming established in new localities into which they are artificially brought by one or another agency, are the climatic and edaphic limits beyond which a given plant cannot exist. ALPHONSE DE CANDOLLE made a distinction, in cases where plants are transferred to new regions, between the *acclimatization* of plants, when they find themselves where conditions differ markedly from those of their native habitat, and the *naturalization* of plants, when they are brought into a region with habitat conditions similar to those to which they have been accustomed. In case of complete acclimatization or naturalization, a plant, in his opinion, gradually acquires all the properties inherent to wild species, such as the ability to grow and to multiply by natural means of propagation, both sexual and asexual, without man's aid. If such a plant maintains itself for several years, during which there occur climatic phenomena to which it has not been accustomed, we may regard it as fully acclimatized. In other words, such a plant would in no way differ from an indigenous one, and its artificial introduction into an alien flora may be established only on the basis of historical data.

It should be noted, however, that, with the exception of weeds, very few instances are known of the full acclimatization of a plant occurring during the history of man. DE CANDOLLE, as a negative example cites the date palm, which has been known to man for thousands of years and which, despite many attempts to extend its area, cannot be successfully grown beyond the limits of its climatic minimum. Although it may be grown at more northerly latitudes than its native home, *e.g.*, in southern Europe on the shores of the Mediterranean, it either yields no fruit at all or fruits that fail to ripen.

Hence, with a few exceptions, we can speak about the acclimatization of plants only in the sense of the gradual adaptation of a plant to changed habitat conditions over a very prolonged period of time, no doubt exceeding that span during which man has acted as a factor in the geographical distribution of plants. In this connection WILLIS (1922) quite correctly points out that acclimatization does not necessarily involve the transfer of plants from one place to another. During the history of vegetation plants have many times had to become acclimatized *in situ* to the great climatic changes that have occurred during geological ages. WILLIS sums up his chapter on this subject as follows (p. 31): —

"Acclimatisation in the hands of man, who is impatient of results, has been largely a matter of trial and error, with numerous failures, but there is reason to suppose that this is not so much the case in the hands of nature, working as she does over vast periods of time, with very small steps. Species have thus been acclimatised to conditions wonderfully different from those in which they began".

In other words, in studying the artificial factors in plant distribution we should keep in mind that these embrace not acclimatization but only naturalization, a first step in acclimatization, this constituting a limitation, and a very real one, to the possibilities of introducing new members into a flora at the will of man.

As regards naturalization, numerous cases are known where this has taken place or is taking place, which indicates, as DE CANDOLLE pointed out, that in the various regions of the earth there do not grow by far all the species which these regions could maintain and which could thrive there. At the same time, however, it is impossible to draw a sharp distinction between acclimatization and naturalization, since, though there are regions with similar climatic conditions, there are no two regions in the world with entirely identical conditions, and, consequently, a plant must pass through a number of long stages in the process of becoming adapted to new conditions before it may be regarded as fully naturalized. These stages of gradual adaptation to new conditions are particularly clear as regards climatic, especially temperature, conditions. As an example, we may take the olive trees (*Olea europaea*) grown in the Crimea. They are the descendants of a number of very old trees preserved from ancient Greek and Genoese settlements. These trees undoubtedly represent cold-resistant races that evolved on the southern coast of the Crimea, the northernmost limit of olive cultivation.

The various stages in a plant's adaptation to low temperature may be illustrated by the acclimatization studies carried out at the Nikita Botanical Gardens in the Crimea (WULFF, 1926). We divided the alien plants there into four different groups according to their stage of naturalization. The first group included plants fully naturalized (in DE CANDOLLE's sense), *i.e.*, capable of self-propagation, growing outside the limits of cultivated fields, and, in general, having acquired the attributes of indigenes. (As examples we may mention: *Daphne laureola*, *Maclura aurantiaca*, *Mahonia aquifolium*, *Cytisus laburnum*, *Bupleurum fruticosum*). In the second group we assembled those species that propagated themselves freely by seeds in the plots where they were cultivated but that did not spread beyond these plots and did not revert to a true wild state. (Examples: cedars, *Trachycarpus excelsus*). The third group, including the majority of the plants in the Nikita Botanical Gardens, embraced those species that fruited normally and produced viable seeds but that were incapable of self-propagation.

Lastly, in the fourth group we placed those species that showed normal vegetative growth but that, due to lack of the proper conditions for their normal development, did not flower at all or, if they flowered, did not form any fruits or formed fruits without seeds or with inviable seeds. (Examples: *Pinus Gerardiana, Pseudotsuga taxifolia, Clematis Jacqmannii, Pueraria Thunbergiana*).

According to CHERNOVA (1939), out of a total of 681 woody, ornamental species now grown in the Nikita Botanical Gardens only 40 are self-propagating. Most of the others (381) produce viable seed but are not self-propagating, 81 do not produce viable seed, 76 bear fruit but produce no seed, and 103 flower but do not fruit. It is, thus, clear that a plant, even when transferred to a region with similar habitat conditions, has serious obstacles to overcome before it can become fully naturalized. This accounts for the very small percentage of species that actually enter into the composition of the wild flora of a country, not merely here and there crossing the boundaries of botanical gardens or plantations but actually extending their range outside these boundaries. Despite the centuries-long work of botanical gardens and acclimatization nurseries, involving the introduction into cultivation of tens of thousands of species, most of these species, though continuing to this day to grow in these gardens and nurseries under man's protection and with his care, have not in the slightest been able to emancipate themselves from this dependence. Any number of examples might be cited to illustrate this. Thus, though thousands of species of woody and herbaceous plants have been introduced at the Nikita Botanical Gardens, only about 700, almost exclusively woody species, have been preserved to the present day. The records show that there existed at these Gardens considerable collections, several times replenished, of herbaceous plants, but of these not a trace has been preserved.

STANKOV (1925) made a count of the species in southern Crimea that had spread beyond the limits of the parks, where they had at one time been grown, and had become wild. Of such species he found that he could list with certainty only 19. Moreover, the habitats of these plants, with very few exceptions, were found to be near the places where they had once been cultivated. These plants are not widely distributed, and it is evident, beyond any doubt, that they are escapes from cultivation.

Though cases are not rare of accidentally or intentionally introduced species spreading very rapidly and, on territory where the native vegetation has been disturbed or destroyed, acquiring a dominant position, it usually happens that such species later as suddenly disappear as they had appeared. Thus, FLAHAULT (1899) describes how in 1856 there were listed at the Montpellier Botanical Gardens 24 naturalized species, all of which later ceased to propagate themselves independently. In 1893 in the same gardens there were 57 flowering species alien to the local flora distributed throughout the territory of the gardens and constituting noxious weeds that had to be controlled. Six years later, in 1899, seven of these species had completely disappeared, and four had ceased to be universally distributed. Taking their place, nine new species had made their appearance. But not one of these plants ever spread beyond the boundaries of the gardens.

Kupffer (1925) points out that the Baltic ports in former times were regions of the concentration of adventive plants brought by ships together with their ballast. In more recent times, the loading of ballast on ships having been entirely discontinued, all this flora has disappeared, with the exception of a very few species, such as *Diplotaxis muralis* and *D. tenuifolia*. Its place was then taken by a different adventive vegetation distributed along the railways, at grain elevators, etc. During and after the first World War the through transit of freight from Russia into the Baltic countries almost entirely ceased, causing the disappearance of many species from this vegetation. During the war years, 1915–1917, there made their appearance very many new adventive plants brought together with provender from southern Russia. After a few years these newcomers likewise almost completely disappeared, being crowded out by species of the local flora, that again took possession of their former habitats as soon as they were no longer so trodden down by great masses of men and horses.

These phenomena of the sudden disappearance of adventive species that had already apparently acquired full rights of citizenship, if not within the local flora, in any case within the limits of the territory cultivated by man or surrounding his dwelling places, may be ascribed to the incomplete naturalization of these plants. The occurrence of an extreme variant of any one of the local habitat conditions suffices to destroy these plants altogether or to injure them to such an extent that they lose out to their competitors in the struggle for existence.

How diverse are the habitat requirements of a plant and how difficult it is for us at once to understand the reason for its incomplete naturalization may be seen from the following two examples of sterility in cultivated plants reported by Cammerloher (1927). *Thunbergia grandiflora* (of the family *Acanthaceae*) grows wild in India, where it propagates itself freely. In Java, on the other hand, where it was introduced because of its large, beautiful flowers, it never fruited, despite the presence of the necessary pollinators. Artificial pollination also proved of no avail This seemingly inexplicable phenomenon found its explanation in the fact that all the specimens of this plant growing in Java had arisen by vegetative propagation from a single plant brought from India and, consequently, constituted one single "clone". The absence of fruits was, therefore, due to the self-sterility of this clone, which was proved by the introduction of a new specimen from India, whose pollen induced the development of the ovaries of the flowers of the old, hitherto fruitless specimens. As a second example let us take *Aristolochia arborea*, three specimens of which were brought from Mexico to the Buitenzorg Botanical Gardens (Java). In their new home these trees flower abundantly during almost the entire year, but they fail to bear any fruit. There is no lack of pollinators, and artificial pollination is fully successful. The cause of sterility in this case has proved to be the fact that for the biological development of the given plant the climate should not be too humid. The pistils and stamens complete their development at different times, the former a day before the latter. Observations in Java of the flowers at the time of opening showed that, when first open, the flowers are already in the second (staminate) stage of their development, while at the pistillate stage they are always closed. This is ascribed to the fact that the climate in Java is more humid than in Mexico, the home of this species, this greater humidity retarding the opening of the flowers by one day. Consequently, at the moment of complete development of the stigmas the flower is still closed and insects cannot penetrate therein, while by the time the flower opens the stigmas are already wilted and have lost their receptivity.

These two examples suffice to show how varied and precise are the biological adaptations of plants to sundry habitat conditions (very often not perceptible to us), any slight alteration in which may make the plant incapable of a normal, independent existence. If to this we add the unceasing struggle for existence and the competition with the indigenous vegetation, we should not be surprised at the relatively small number of those species introduced by man for cultivation or accompanying him in his migrations that become fully naturalized components of the local flora.

This explains, for instance, the fact that the numerous species dispersed to distant regions by railways very rarely spread beyond the roadbed itself. Thus, LITVINOV (1926) reported the finding of southern plants along the Murmansk Railway. Some of them grew there year after year for a considerable number of years, but in no case did they spread beyond the immediate vicinity of the railway line and stations. The fact that these southern plants could maintain themselves so far north over a period of years must undoubtedly be accounted for by the presence of sand sprinkled along the roadbed and by the absence of competition, that potent factor in the life of organisms.

Similarly, the flora of Montpellier has been enriched to only a very slight extent by the numerous species brought in with the wool at Port Juvenal. According to THELLUNG (1912), the number of species that have been introduced in various ways by man into the flora of Montpellier beginning with the sixteenth century is not over 107, or 3.8 per cent of the total number of species in this flora, estimated by him to be 2,792. The smallness of this percentage may be ascribed largely to competition with the local flora and the latter's full occupation of the available territory, factors whose importance we have already stressed.

Adventive plants are ordinarily able to spread only in places where the natural vegetation has been destroyed by man and the habitat conditions to which the plant communities formerly dwelling on this territory had been accustomed have been altered. Especially favorable conditions are created for the adventive plants, when man, cultivating his land year after year, does not allow the local vegetation to regain its hold. But once such land is left for a time uncultivated, there ensues a struggle between the aliens and the indigenes that usually ends in the victory of the latter. It is precisely the destruction of the indigenous flora and the marked changes in habitat conditions caused by man that make possible the rapid and widespread distribution of adventive vegetation on islands and on those territories of the mainland newly utilized and colonized by man, for it is in such places that natural plant communities are destroyed prior to cultivation of the land. This holds particularly true in those cases where forests are cut down and the newly introduced vegetation receives a habitat with rich, unexhausted soil and without competitors. But even in such cases the invaders ordinarily maintain their ascendancy only under the protection of man, who does not allow the indigenous vegetation to regain its usurped rights. But once man deserts any such place of habitation, the indigenous vegetation with amazing rapidity returns to its own, this being particularly striking in forest regions, where abandoned cultivated land and settlements are again so densely covered with woods as to be unrecognizable.

Views formerly held that introduced species (aliens) may completely crowd out the indigenous vegetation have, after detailed studies of the resultant interrelations, been discarded as erroneous. Thus, ALLAN (1936) points out that in the flora of New Zealand out of about 600 aliens only 48 may be regarded as serious competitors of the indigenes. These 48 include 28 from the Old World, 7 from Australia, 9 from the Americas, and 4 from South Africa. These species have a localized distribution limited to formations modified by man and under his constant protection, *e.g.*, pastures. If such vegetation had been compelled to develop independently, without man's protection, the indigenes would, undoubtedly, have crowded out the aliens. ALLAN concludes: "We are left with the result that only some half dozen aliens can truly be said to have suppressed any indigene—and that very locally" (p. 191).

Similar and extremely interesting data have been reported by PERRIER DE LA BÂTHIE (1932) for Madagascar. The total number of species that have been introduced into Madagascar accidentally or intentionally by man is 524. To this number we must add 380 species alien to the local flora, which reached the island as a result of natural factors of dispersal or the manner of whose entry is unknown but which are distinguished by the same characters as the flora introduced by man: absence in indigenous plant communities, occurrence in open associations or in those modified by man, possession of characters alien to the local flora, high generic coefficient (ratio of number of genera to number of species), seeds whose wide dissemination is readily possible. Consequently, the total number of species alien to the flora of Madagascar is approximately 900, a very large number, but this is not surprising when we recall that about 70 per cent of the natural vegetation of Madagascar has been destroyed by man. Of all these 900 species, however, only one, *Adenostemma viscosum*, has gained a foothold in natural plant communities undisturbed by man. This is an exceptionally remarkable and important fact.

As long ago as 1899 FLAHAULT emphasized the fact that the boundaries of the area of a species depend on environmental conditions to a far greater degree than had been supposed and that the slightest deviation in these conditions may threaten the existence of a species, may even cause it to suffer annihilation in the struggle for existence. This explains why in many cases adventive vegetation, apparently fully naturalized, ceases to bear fruit and to spread independently as soon as man leaves it to itself and deprives it of his direct or indirect protection.

Geographical Limits to the Distribution of Exotics: — There are a number of exotics seemingly unaffected by the above-described obstacles to their dispersal, there apparently being no limit to their spread over the globe. Among such plants found in Europe, seemingly independent of the laws of the geographical distribution of plants, we may mention: *Elodea canadensis*, *Amaranthus retroflexus*, *Impatiens parviflora*, *Veronica Tournefortii*, *V. polita*, *V. agrestis*, *Linaria cymbalaria*, *Solidago serotina*, *Erigeron canadensis*, *Xanthium spinosum*, *Galinsoga parviflora*, and several American species of asters. But more thorough and detailed study shows that even in the case of such cosmopolitan species, despite the exceptionally wide range of their adaptability, their distribution has its geographical limits.

Dr. A. I. Maltsev, in his monograph on the weeds of the U.S.S.R. (1926), has assembled considerable data on the geographical distribution of weeds and exotics. On the basis of many years of investigation, Dr. Maltsev concludes that only a very few weed species, such as *Agropyrum repens*, *Capsella bursa-pastoris*, *Chenopodium album*, and *Taraxacum officinale*, have extensive areas of distribution, these species being characterized by numerous forms and races adapted to diverse habitat conditions. In the case, for instance, of *Taraxacum officinale*, a monographic study of the species has shown it to be a coenospecies and has shown the need of dividing it not only into various races but into several, separate species. It may be that this will also be found to hold true for the other species we have just enumerated.

Most weeds have more limited areas of distribution, beyond whose bounds they cannot pass. Thus, wild oats, *Avena fatua* ssp. *septentrionalis*, is distributed only in the northern part of European U.S.S.R., while in the south it is replaced by another form, *Avena fatua* ssp. *fatua*. Many weeds are confined chiefly to the Black Soil zone, beyond whose limits they rarely extend (*e.g.*, *Allium rotundum*, *Falcaria Rivini*, *Vaccaria parviflora*, and others); outside of this zone they appear again only in the Crimea and the Caucasus. Very common weeds in central regions of the U.S.S.R., such as *Cirsium arvense*, *Sonchus arvensis*, and *Agrostemma githago*, are replaced in the southeast by entirely different weeds. The extreme south has its own weeds, such as *Centaurea picris*, various species of *Caucalis*, etc. Weeds of the forest zone in passing to steppe regions often lose their weed character. Some weeds, such as *Centaurea cyanus*, known to ancient writers, on spreading south, even show a break in their distribution, disappearing in the arid, steppe zone and reappearing in the Crimea and the Caucasus (Pachosky, 1911; Koroleva, 1930).

Conclusions: — The significance of man in changing the vegetation of the earth, particularly as regards disturbing its natural state, is unquestionably immense; his rôle in the distribution of plants over the globe is, no doubt, very considerable, but it is, nevertheless, greatly limited by the specific habitat conditions required by each particular species of plant. The significance of man's rôle with respect to the present distribution of species throughout the world has been greatly exaggerated, and it is only during the past few centuries that man's rôle has acquired the importance that it has. Only the comparatively recent, extensive development of ways of communication and the constant movement of people and freight from region to region, as well as the expansion of the cultivated areas and the introduction into cultivation of plants from distant countries, have enabled man to assume such an important rôle in the geographical distribution of plants. It is entirely clear that his significance in bygone centuries becomes less and less. Consequently, there can be no doubt that for an explanation of complex problems in historical plant geography, for an understanding of the present areas of ancient species of the plant kingdom, and for the solution of problems connected with discontinuous areas of plants, the activity of man is a factor of altogether too recent date. In all those cases where it can be established that any given changes in the character of an area were induced by man's activity, their relatively recent nature may be taken for granted.

References:

ALLAN, H. H., 1936: Indigene versus alien in the New Zealand plant world (Ecology, Vol. 17, pp. 187–193).

ALLAN, H. H., 1937: The origin and distribution of the naturalized plants of New Zealand (Proc. Linn. Soc. London, Session 150, Pt. I, pp. 25–46).

BATALIN, A., 1892: Das Perenniren des Roggens (Acta Horti Petrop., Vol. 11, No 2).

BRAUN-BLANQUET, J., 1923: L'origine et le développement des flores dans le massif central de France (Paris).

CAMMERLOHER, H., 1927: Javanische Studien, I. Über einige Fälle von Unfruchtbarkeit kultivierter Pflanzen fremder Florengebiete (Oest. Bot. Zeitschr., Vol. 86, No. 1).

DE CANDOLLE, A., 1855: Géographie botanique raisonnée, Vol. II, Chap. 8 (Paris).

CHERNOVA, N. M., 1939: Brief summary of experimental work with trees and shrubs at the Nikita Botanical Gardens (In Russian; Bull. Nik. Bot. Gard., Vol. 22, No. 1).

COCKAYNE, L., SIMPSON, G., and THOMSON, J. SCOTT, 1932: Some New Zealand indigenous-induced weeds and indigenous-induced modified and mixed plant-communities (Journ. Linn. Soc., Bot., Vol. 49, pp. 13–45).

FLAHAULT, M., 1899: La naturalisation et les plantes naturalisées en France (Bull. Soc. Bot. de France, 3 sér., Vol. 6).

HILDEBRAND, FR., 1873: Die Verbreitungsmittel d. Pflanzen (Leipzig).

KOROLEVA, A., 1930: Survey of literature on the blue cornflower, *Centaurea cyanus* L. (In Russian; Bull. Appl. Bot., Gen., and Plantbr., Vol. 22, No. 5).

KUPFFER, K., 1925: Grundzüge der Pflanzengeographie des Ostbaltischen Gebietes (Abh. d. Herder Inst. zu Riga, Vol. 1, No. 6).

MALEYEV, V. P., 1933: Theoretical bases of acclimatization (In Russian, English summary; Bull. Appl. Bot., Gen. and Plantbr., Suppl. 60).

MALTSEV, A. I., 1926: Weeds of the U.S.S.R. and their Control (In Russian; Leningrad).

MALTSEV, A. I., 1929: Latest achievements in the study of weeds in the U.S.S.R. (*In* "Achievements and Prospects in the Field of Applied Botany"; in Russian; Leningrad).

PACHOSKY, I., 1911: On the Field Weeds of Kherson Province (In Russian; Kherson).

PERRIER DE LA BÂTHIE, H., 1932: Les plantes introduites à Madagascar (Rev. Bot. Appl., Vol. 12).

RIKLI, M., 1902: Antropochoren und der Formenkreis des *Nasturtium palustre* (Ber. Schweiz. Bot. Ges., Vol. 13).

STANKOV, S., 1925: On some characteristic cultivated and naturalized plants of the southern coast of the Crimea (In Russian, English summary; Bull. Appl. Bot., Gen., and Plantbr. Vol. 14, No. 4).

THELLUNG, A., 1912: La flore adventice de Montpellier (Mitteil. d. Bot. Mus. Zürich, Vol. 58).

THELLUNG, A., 1915: Pflanzenwanderungen unter dem Einfluss des Menschen (Bericht d. Vereinig. für Pflanzengeographie).

THELLUNG, A., 1930: Die Entstehung der Kulturpflanzen (München).

WILLIS, J., 1922: Age and Area, Chap. IV. Acclimatisation (Cambridge).

WULFF, E., 1925: Contribution to the history of the experimental work conducted at the Nikita Botanical Garden during the period 1813–1860 (In Russian; Bull. Nik. Bot. Gard., Vol. 8).

WULFF, E., 1926: Der Nikitsky Botanische Garten in der Krim (Mitteil. d. Deutsch. Dendrol. Gesellschaft, Vol. 37).

ZINGER, N., 1909: On species of *Camelina* and *Spergula* infesting flax and their origin (In Russian; Bull. Bot. Museum Acad. Sci., Vol. 6).

Chapter VIII

NATURAL FACTORS IN THE GEOGRAPHICAL DISTRIBUTION OF PLANTS

In undertaking a study of the geographical distribution of plants, one is, first of all, confronted by the question: Is the dispersal of a species mass-like in character, the species as a whole gradually, step by step, though perhaps unevenly, covering ever new territory until some barrier puts a limit to its spread, or is its dispersal achieved by sudden spurts, as a result of individual plants penetrating into distant localities isolated from the main area? In the first case discontinuous areas could result only from the dying out of the species in the intervening territories or the isolation of the occupied regions due to geological causes, and such areas would, consequently, have to be regarded as relic. In the second case the cause of their discontinuity would be the chance transport of plants or parts of plants, such as seeds, a great distance from the mother plants.

We know that plants possess various structural adaptations for the dissemination of their spores, seeds, and fruits, and that the latter, by means of these adaptations supplemented by various natural factors, such as wind, sea and river currents, animals, birds, etc., may be transported a considerable distance from their place of origin. In this connection, however, another question arises: How far may seeds be carried and still retain their germinating power? We have already noted that there exists a limit in this respect, differing, of course, for different species. In order to arrive at some conclusions and to give an answer to the questions we have just put, we shall now take up the natural factors which facilitate the dispersal of plants.

Wind as a Factor in the Dispersal of Plants: — Atmospheric movements induced by various air currents—especially the stronger movements, winds—doubtless facilitate the dissemination of spores and seeds for considerable distances from the place of their formation. The lighter atmospheric currents are naturally of significance only for the transfer of very light objects, which, by reason of their lightness, remain in the air, as if suspended, and move laterally and also vertically at the slightest movement of the air.

In the dispersal of cryptogams, their spores being exceptionally light, even slight air currents may play a big rôle. FALCK (1927) found that the spores of *Basidiomycetes* and particularly of *Ascomycetes*, owing to the turgor of the cell-sap at the time of maturation, are forcibly ejected from the ascus, and that, even when the air is practically motionless, these spores are borne upward due to atmospheric currents induced by temperature conditions. Subsequently, winds and, in general, the stronger horizontal movements of the air aid in carrying them over considerable distances.

Members of the staff of the U. S. Department of Agriculture

(STAKMAN *et al.*, 1923) made some interesting observations from airplanes as to the number of spores in the upper layers of air. The spores were caught on ordinary microscope slides smeared lightly with vaseline on one side. The slides were exposed for a definite period of time, after which they were returned to the slide box to be analyzed later in the laboratory. The observations were made in Texas in 1921 in different months of the year, and it was found that up to an altitude of 11,000 feet above the surface of the earth the atmosphere contains numerous spores and pollen grains, as many as 10 being caught on a slide exposed for 5 minutes, while still higher (observations were made at altitudes as high as 16,500 feet) no more that one or two were caught on a slide exposed for 10 minutes. At low altitudes, say 1,000 feet above the surface of the earth, as many as 450 were caught on slides exposed for 20 or 30 minutes.

For flowering plants such slight atmospheric currents cannot be of as much significance, since their seeds, even the very smallest, are too heavy to remain suspended in the air. Stronger air currents are needed for their dispersal, it being understood, of course, that the distance seeds may be transported by wind from the place of their origin depends on the velocity of the wind, on the size and weight of the seeds, on whether or not they possess special adaptations for flight, and on the effectiveness of these adaptations.

The transport by wind not only of seeds but even of small animals and other objects for considerable distances has been observed on numerous occasions. We shall cite a few instances, taken, for the most part, from a review of the literature on this subject made by BÉGUINOT (1912). VISIANI described the transport by wind into Mesopotamia of considerable quantities of various parts of *Lecanora esculenta;* KOLDWEG found leaves off the coast of Greenland 15 km. from the shore; SVERDRUP found leaves of cereals on the polar ice at a great distance from any land. VOGLER (1901) reports several cases of the transport of leaves into the upper altitudes of the Swiss Alps, considerably higher than the forest line. Thus, beech leaves were found at an altitude of 2,490 m., 10 km. from the nearest habitat of beeches, and leaves of *Alnus viridis*, *Sorbus aria*, willows, etc. at an altitude of 2,585 m., 15–20 km. from the nearest trees. In the Alps on the Franco-Swiss boundary (on the summit of Les Cornettes de Bise) BEAUVERD observed on the snow at an altitude of 2,439 m. beech leaves and maple samaras, the nearest habitat of such trees being 4 km. away and about 1,000 m. lower. MIRBEL reports the wind-transport into Spain of seeds of plants growing on the northern shore of Africa. WALLACE tells of large quantities of seeds of some unknown species of *Salix* being carried into Shanghai by the wind. WARMING found on the eastern shore of Jutland faded flowers and fruits of species of *Calluna* and *Erica*, blown by the wind across the Cattegat from Sweden.

VOGLER (1901) devoted a special paper to the problem of the significance of wind in the distribution of alpine vegetation. From his statistical study of the flora of Switzerland he found that among true alpine species 59.5 per cent possess adaptations for the wind-dispersal of their seeds or fruits, while among non-alpine species this percentage is only 37.9. Another interesting conclusion made on the basis of the data of this study is that in the Alps the higher the altitude, the lower the percentage of species possessing adaptations for the dissemination of their seeds or fruits by animals and the higher the percentage of wind-dispersed species.

These data fully suffice to assure us that seeds and fruits are dispersed by the wind and that by strong winds they may be borne for a considerable distance from the place of their formation. This conclusion is beyond dispute. Other questions, however, arise: To what extent are such seeds capable of germinating and taking root under the new conditions? Over what distances may seeds be borne by wind? What significance does wind have as a factor in the "saltatory" dispersal of plants (dispersal to widely separated localities), *i.e.*, as a factor in the discontinuous distribution of plants? Unfortunately, we have very few concrete data as to the transport of seeds and fruits by wind, and indirect proofs or data that are not exact are always subject to dispute. This accounts for the wide divergence of opinion with respect to the evaluation of the significance of wind in the transport of seeds over great distances.

Among investigators ascribing great significance to wind in the dispersal of seeds and considering that discontinuous areas may arise as a result of the transport of seeds by wind is VOGLER, who in the work already cited lists twelve concrete facts of such dispersal. According to data presented by him, seeds have been borne by the wind over mountain peaks more than 2,000 m. in height; horizontally, the maximum distances noted by him for the wind-transport of seeds are: 25 km. for *Cardamine resedifolia*, 9 km. for *Galeopsis speciosa*, 3 km. for *Serratula rhaponticum* and *Arabis turrita*. Despite the great significance ascribed by VOGLER to wind in the dispersal of plants, we find even in his own conclusions small grounds for his belief in the important rôle of wind in the origin of discontinuous areas. Thus, he concludes (p. 75) that the transport of seeds by wind for great distances, even for hundreds of kilometers, is possible, but practically such long-distance wind-transport plays a very limited rôle in the dispersal of plants; of great significance is the wind-transport of seeds up to 40 km. and also such transport over high mountain ranges.

VOGLER's work, therefore, as he himself points out, proved irrefutably the significance of wind as a factor in the *gradual* dispersal of plants, particularly as regards the invasion of territory with land as yet unoccupied by vegetation, such as inaccessible cliffs and canyons. However, as a proof of the "saltatory" dispersal of plants, as an aid to an understanding of discontinuous areas, it contributed nothing, since the isolated habitats mentioned by him either are located at relatively short distances from one another, or are clearly of a chance nature, or may be fully explained by the local topography of the area.

Let us now examine the arguments of those investigators who consider that the wind plays an insignificant rôle in the dissemination of seeds over great distances. ALPHONSE DE CANDOLLE, in his "Géographie Botanique", pointed out that wind is often a factor in the dispersal of plants to the extent of transporting their seeds within the limits of their habitats or to adjacent territories. In many cases the seeds alight in places where their growth and further development is impossible, in consequence whereof they are doomed to perish, but sometimes, due to a sudden change in conditions, seeds accumulated perhaps over a period of years and lying dormant suddenly find it possible to develop. This may occur, for example, when a forest is cut down, when there may be observed the unexpected development of a number of species that formerly could not grow in this forest but whose seeds were borne each year into the forest and to a partial extent remained inviolate and preserved their ability to germinate. In mountainous regions the wind carries seeds over hills and mountain passes, wafting them to considerable heights, whence they fall, however, at not a great distance from the mother plants. Thus, according to HUMBOLDT, BOUSSINGAULT saw seeds lifted by the wind to a height of 5,400 m., whence they again fell not far from their original location.

But, notwithstanding such irrefutable facts, DE CANDOLLE doubts that seeds are borne by the wind over great distances. He never heard, for instance, of seeds being carried by winds from England to France or from Ireland to England; he never had proof of the transport by wind of seeds from Africa to Sardinia, from Sardinia to Corsica, or from Corsica to the shores of the mainland near Genoa or Nice, despite the force of the south winds prevailing there. He bases his conclusions also on other authorities on the problem of the dispersal of plants, such as GUSSONE, GODRON, and LYELL, and also on the testimony of sailors, who observed the transport by wind of winged insects or dust to the decks of ships but never saw seeds. Hurricanes, such as rage, for example, over the Antilles or along the Chinese coast, must, of course, also carry with them numerous seeds and even fruits, but these hurricanes are, for the most part, purely local in character and revolve in a circle whose diameter is not very large. An important point made by DE CANDOLLE is that wind can be of great significance in seed dispersal only in case of their mass accumulation, which never occurs.

KERNER (1871; 1891) noted that on sunny, windless days numerous plumed seeds and fruits are borne aloft to a considerable height by ascending air currents. He tried to determine the number of such seeds and counted during one minute as many as 280 seeds floating past him and rising in a current of air, so that it may be presumed that the total number of such seeds borne upward by atmospheric currents during a single day amounts to several millions. Nevertheless, for the dispersal of plants over great distances this has no significance, since after sunset these seeds and fruits again alight not far from the place whence they were lifted. KERNER further points out that in some plants the special adaptations for flight remain attached to the fruits or seeds only during the first flight; once the latter have alighted, these adaptations fall off. This is true particularly in the case of pine seeds, which after they have once alighted are no longer able to fly; it also holds true in the case of many *Compositae*, whose plumed achenes, drifting with the wind, at the slightest collision with some obstacle become detached from their plumes and fall to the ground.

RIDLEY (1930, pp. 131–2) points out that, in the case of plumed fruits, in measuring the distance to which they may be carried by the wind, it is necessary to distinguish between the transport of the downy plumes alone and the transport of plumes still bearing the seeds or achenes. Records of distant flights refer chiefly to the plumes alone; as to how far they flew with the fertile achenes still attached data are entirely lacking.

ILYINSKY (1933) thus records his observations on the wind-dispersal of spruce seeds (*Picea excelsa*) in the upper Volga Basin: "Seeds once fallen to the ground did not move any farther at all or made but very short hops, the entire distance between them and the mother tree being measured at the most in tens of meters." According to SCHMIDT (1918), the distance to which seeds and fruits provided with special adaptations for flight may be carried by the wind is not more than a few kilometers and, in some cases, not over a few meters, *e.g.*, the birch—1.6 km., the maple—0.09 km., and the ash—0.02 km.

BENTHAM (1873) points out that the rôle of wind-borne seeds in the dispersal of species of *Compositae*, according to data assembled by him, is less than in the *Leguminosae*, *Labiatae*, and other families not possessing any adaptations for seed dissemination. His investigations showed that only 60 species of this family, *i.e.*, not over 0.67 per cent of the entire number of its species, are characterized by "saltatory" dispersal by whatever agency, including accidental transport by man, while the other 99.33 per cent of the species either remained within the limits of their initial areas or spread farther by a gradual extension of their area.

Discussing the wind as a factor in the geographical dispersal of seeds, particularly of the *Compositae*, BENTHAM confirms the conclusion of DE CANDOLLE and KERNER as to the inability of plumes to bear fruits over great distances, especially over the sea. A very strong wind cannot carry them farther than two or three miles, due to the fact that, once the plumes encounter humid atmosphere, they become stuck together, causing the fruits to fall to the ground, whence they are only very rarely borne farther by the wind. Thus, plumes can transport seeds only short distances from the mother plants, and the only way in which they might facilitate the migration of species is by accidentally aiding the seeds being blown into streams, bales of goods, or soil carried as ballast, whence they might be carried much farther. That the rôle of plumes in plant migration is of slight significance is indicated also by the fact that of the 60 species mentioned above as characterized by "saltatory" dispersal only 22 or 23, or about one-third, have plumes, whereas over two-thirds of the species of *Compositae* are provided with them.

RIDLEY (1930; pp. 71 and 130), despite his strong tendency to ascribe to natural factors and particularly the wind very great significance in the dispersal of plants, writes that winged fruits and seeds are carried by the wind for only a relatively short distance, "not more than 1 mile or so".

On the basis of the foregoing data we feel fully justified in concluding that the wind is a very important factor in the dissemination of seeds over short distances but has little significance in their transport over long distances, and that, consequently, most discontinuous areas cannot be explained by its action.

Water as a Factor in the Dispersal of Plants: — Adaptations for the dispersal of plants by water are possessed chiefly by plants living in or near water; nevertheless, due to the ability to float of most fruits, seeds, and other parts of plants, it is quite probable that they may be dispersed by water. We should, however, draw a clear distinction between the dispersal of aquatic and terrestrial plants. As regards the former, their dispersal by means of water—sea-currents, rivers, etc.—is obvious and does not require any special proof. With respect to terrestrial plants, however, we can only assume the possibility of their dispersal by sea and river currents, floating islands, ice-cakes, etc. We shall consider each of these modes of dispersal separately.

1. *Sea-Currents.* — As early as 1605 CLUSIUS and, somewhat later (1696), SLOANE published papers, in which they directed attention to fruits borne by Atlantic Ocean currents to the shores of Scotland, Ireland, and the Orkney Islands. LINNEAUS wrote about fruits and seeds of plants borne by the Gulf Stream from the American tropics (apparently from the Antilles) to the shores of Norway. Since those times numerous authors have cited analogous facts of the dispersal of seeds, fruits, and other parts of plants, even floating trees, but we need not enumerate them here (for further details see HEMSLEY, 1885, RIDLEY, 1930, and others).

Nevertheless, the question as to whether or not sea-dispersed seeds can germinate, establish themselves, and serve for the further dispersal of the plants was at first scarcely touched upon. One of the first to give it serious attention was DARWIN. He made a number of experiments which showed that seeds of many plants after prolonged immersion in sea water preserve their ability to germinate. Despite the favorable data that he obtained from these experiments (which, it should be noted, were quite primitive and, hence, not very convincing), his conclusions on this point are far from indicating definite conviction even on his part. As he puts it, "The floras of distant continents would not by such means become mingled; but would remain as distinct as they now are" ("Origin of Species", 6th ed., 1911, p. 514).

Interesting experiments were also made by G. THURET (cited by DE CANDOLLE) on the ability of seeds to preserve their germinating power after prolonged immersion in sea water. Out of the many seeds that he tested only those of ten plants survived and preserved their ability to germinate. THURET, and also DE CANDOLLE, came to the conclusion that, with the exception of plants accustomed to grow on sandy shores, it is difficult to presume that seeds borne by sea currents could find suitable conditions for their development and multiplication.

HEMSLEY (1885, p. 42) lists the seeds of over 120 species of plants (belonging to 48 families) dispersed by sea currents. In particular, with respect to the Bermuda Islands, he considers that 45 species of their flora are of such origin. But, despite the undoubted significance of sea currents in the dissemination of seeds, especially as regards the stocking of islands devoid of vegetation, such as coral islands, there are facts showing that in some cases these currents have not played the rôle as a factor in the geographical distribution of plants attributed to them. As an example we may take the coco palm, a typical littoral plant. Its wide range within the tropical zone and the ability of its fruits to protect its seeds for a long time from injury by sea water would suggest that sea currents probably played an important rôle in the dispersal of this palm. This is supported by the fact, for instance, that great numbers of these palms were already growing on the Cocos-Keeling Islands, in the Indian Ocean, at the time of man's first settlement there, in 1827. It is strange, on the other hand, that the coco palm, which grows on the Pacific coast of Central America, is not found on the nearby Galapagos Islands, which lie in the very path of the sea currents passing from the shores of America and bearing various seeds to the shores of these islands. Likewise there are few or no coco palms on the Molucca Islands. These facts have led most investigators to believe that, although coconuts are doubtless transported by sea currents, this is not sufficient to insure that their seeds take root in their new abode. Coconuts simply lying on the surface of the soil seldom germinate; in order to germinate, they need to be imbedded in the soil. Consequently, the distribution of this palm on certain islands of the Pacific Ocean must have been due to their having been originally planted there by man (HEMSLEY, *l.c.*, p. 306). This latter point of view is upheld also by COOK (1901), who cites a number of striking facts showing that the coco palm could not possibly owe its distribution to dispersal by sea currents. BECCARI (1916), on the other hand, believes that sea-dispersal of the coco palm is not beyond the range of possibility.

Another interesting example of the fact that the mere transfer of seeds to new places is inadequate to establish with certainty that the seeds germinated and that the area of the given plant was thus enlarged is *Lodoicea sechellarum*, which grows exclusively on the shores of the Seychelles Islands. Prior to 1743, when these islands were discovered, only the enormous nuts of this palm were known, found floating in the sea or on the shores of India and of the Malay Archipelago, particularly the Maldives (HOOKER, 1827). Despite the fact that the fruits of this palm have been carried such great distances by sea currents, it grows nowhere except on the Seychelles Islands.

Among various other investigations of this question we wish to mention those of SCHIMPER (1908) on the adaptations of fruits for floating and the resistance to the action of sea water possessed by seeds of plants inhabiting the shores of the islands of the Malay Archipelago. On the basis of his observations he listed about 100 species of sea-dispersed plants. From his study of mangroves growing on the shores of tropical Africa he found that the species composition on the eastern and western shores was entirely different. All species of the western shore are found also in the West Indies, while the mangroves of the eastern shore comprise elements from the East Indies. This circumstance he ascribes to the strong sea currents passing these shores.

HÖCK (1901), having studied the distribution of littoral plants along the shores of northern Germany, divides them into three groups and arrives at the conclusion that the plants in the first of these groups, including exclusively littoral plants, were dispersed along these shores when the distribution of land and sea was the same as at present. As regards the other two groups, steppe plants and arctic-alpine plants, in order to explain their present distribution along these shores, he finds it necessary to seek historical causes and to presume the existence, at the time of their invasion of the region, of a different distribution of land and sea and of different climatic conditions than those now prevailing.

RIDLEY (1930) notes that "nearly all sea-dispersed plants belong to genera which have a large number of terrestrial inland species, and that, as a rule, there are only one or two species which are adapted for sea-dispersal" (p. 250).

2. *Rivers and Streams.* — The transport of fruits, seeds, and other parts of plants by rivers and streams, flowing from mountains and hills into valleys and plains, is well known. Instances are not uncommon of fruits and seeds that have been borne by rivers for exceedingly great distances—from the mountains down to the very sea. BÉGUINOT (1912) found, over a territory two kilometers in extent, among the debris washed down to the mouth of the Tiber River cyclamen tubers and seeds of as many as twenty-five different species of plants.

Rivers undoubtedly also play an important rôle in the seeding of inundated meadows and, in general, of flood plains at times of high water. KELLER (1922), on the basis of a study of the débris left by the Voronezh River (southeast of Moscow) after the spring floods, recorded a considerable number of river-borne fruits and seeds, *e.g.*, the fruits of *Aristolochia clematitis, Iris pseudacorus, Rumex crispus*, several species of sedges, etc. An anatomical study of these fruits showed that they possessed various adaptations for dispersal by floating.

3. *Glaciers, Icebergs, and Ice-Cakes.* — The transport of large stones, boulders, and clumps of earth by glaciers and floating ice-cakes has long been known. This led LYELL to put the question as to the possibility of the chance transport, in this way, of seeds, a mode of transport that might presumably have been of widespread significance during the Ice Age. This view was at first very widely held, it being considered possible, for instance, to explain in this way the stocking of oceanic islands located far from the mainland.

At the present time, the rôle of ice as a factor in the geographical distribution of plants is regarded with considerable skepticism, despite the fact that glaciers may, of course, carry seeds from mountains into the valleys at their feet. This was formerly taken as grounds for explaining the occurrence of mountain plants in far-distant valley habitats as the result of their seeds having been transported there by glaciers during the Ice Age. The supposition of CHRIST (1882) that during the Ice Age mosses and certain ferns (*e.g., Asplenium septentrionale*) were dispersed by means of erratic boulders, based on the fact that these plants are now found on such boulders, was shown by AMANN (1894) to be untenable. The latter investigator studied the various species of mosses growing on such boulders, and he found that they comprise chiefly those species which avoid a lime substratum and which, therefore, have settled on these granite boulders as being the only habitats, due to the limy character of the mountainsides in the Alps and the Juras, that correspond to their biological needs. Moreover, the data on the distribution of these species show that among the species dwelling on erratic boulders there is not one real alpine, high-mountain species and only a very few subalpine species and these such as are found also at lower elevations, while there are a considerable number of species that are found at all elevations and, hence, are not capable of furnishing any data that might give a clue to their origin. Consequently, AMANN came to the conclusion that the mosses found on erratic boulders in Switzerland cannot serve as proof of the transport by glaciers of the seeds of alpine plants from mountains to plains during the Ice Age. It is much more probable that they invaded the plains later during the course of the present geological period.

PETTERSSON (1929) investigated the plants growing on erratic blocks (boulders) near Helsingfors, and he came to the conclusion that these plants belong to the surrounding flora. Seeds from this flora, borne thither by wind or other agents, germinated on these huge boulders after the latter had arrived at their present location.

The foregoing suffices to illustrate the undoubted significance of water, in its different manifestations, as a factor in the dispersal of plants (see also GUPPY, 1917) along sea and lake shores and river banks, but it does not provide an adequate basis for establishing the possibility of the transport of plants for any great distance inland. Even in those cases when seeds are borne by sea currents for great distances, this rarely has any significance in the founding of new, isolated habitats of these plants.

ALPHONSE DE CANDOLLE correctly emphasized the fact that, though rivers may bear the seeds and fruits of plants from the Alps into the plains or for considerable distances from north to south or from south to north, they often bring them to regions where they cannot establish themselves. This holds true also for sea currents, which may carry seeds for immense distances. But even if such seeds, despite prolonged immersion in sea water, preserve their germinating power, they play, from a phytogeographical point of view, a very insignificant rôle. Sea currents usually do not flow parallel to the equator, and so often the seeds are borne into regions where the climatic conditions are entirely alien to them, not to mention the fact that one or two seeds are in most cases quite inadequate for a plant to become naturalized and establish itself as a component part of an already existing vegetation.

Animals as a Factor in the Dispersal of Plants: — That animals play an important rôle in the dispersal of plants is beyond doubt. Nevertheless, it is clear that in most cases seeds are borne by animals over comparatively short distances. For an elucidation of the question interesting us—the possibility of explaining discontinuous areas of species by the chance transport of seeds—only a few kinds of animals, chiefly birds and, to some extent, fish (that carry seeds of water plants), are of any significance. Hence, we shall consider primarily the rôle of birds in the dispersal of plants. The significant rôle that birds may play in the dispersal of plants has been pointed out time and again, beginning with THEOPHRASTUS, who noted that the mistletoe (*Viscum album*) is dispersed by birds.

Birds may serve in two ways for the distribution of seeds over great distances: (*1*) evacuation of undigested seeds with their excreta (endozoic dispersal); and (2) transport of seeds by adhesion to the feathers, feet, or other parts of their bodies (epizoic dispersal). As examples of the first type we may take the facts reported by DARWIN in his "Origin of Species". During two months he found in his garden in the excrement of small birds the seeds of twelve different species, some of which later germinated. Nevertheless, as DARWIN points out, he did not once observe a case of "nutritious seeds" passing through the intestines of a bird uninjured. In this connection the duration of time that a bird retains seeds in its crop is of significance. After a bird has devoured a large number of seeds, they only gradually pass into the gizzard, so that some of the seeds lie uninjured in the crop for as much as 12–18 hours. During this time a bird might, if driven by a gale, cover a distance of 500 miles. When birds of prey fall upon such exhausted birds, "the contents of their torn crops might thus readily get scattered" ("Origin of Species," 6th ed., 1911, p. 510). DARWIN also cites (*l.c.*, pp. 511–512) a number of instances of the finding of seeds in cakes of earth adhering to the feet and legs of birds. He even cites such an extraordinary case as that of obtaining from one lump of earth, weighing 6½ oz., taken from the leg of a partridge sent to him, seed that after a lapse of three years produced 82 plants, 12 monocotyledons and 70 dicotyledons, the latter belonging to at least three different species. Despite our confidence in DARWIN, it is very hard to believe that such a large number of seeds could have adhered to the leg of a bird by natural means.

HEMSLEY (1885) was likewise inclined to ascribe much significance to birds as a factor in the dispersal of plants. He denies the possibility of the transport of seeds in this way over great distances, as GUPPY assumes to be possible (5,000–6,000 miles), but he considers it very probable that birds may carry seeds, for example, from one island of the Polynesian group to another. Birds and sea currents together play an important rôle in the stocking of islands, particularly coral islands, which otherwise would be deprived of vegetation and, consequently, would be uninhabitable. The Bermuda Islands, in particular, owe their vegetation, to a considerable degree, to these two factors. HEMSLEY gives a list of 38 species of swamp plants, the seeds of which might be borne by birds together with particles of mud adhering to their feet, and 13 other species of plants with fleshy fruits that might be transported endozoically by birds.

Nevertheless, it is impossible to regard these facts as universal, and there are a considerable number of cases when the isolated habitats of a species may be explained only by the former existence of a land-bridge connecting these widely separated parts of an area. As an example, HEMSLEY cites *Phylica nitida* of the family *Rhamnaceae*, which has an extensive range, embracing a number of widely separated islands, such as Tristan da Cunha, Réunion (Bourbon), Mauritius, Amsterdam, and, perhaps, Madagascar. The character of the distribution and the distance separating one station from another is such as scarcely to give grounds for presuming the chance dispersal of this plant by birds.

KERNER (1898) conducted an extensive series of experiments on feeding 16 species of birds with the seeds of 250 different species of plants. He found that the birds tested might be divided into three groups:

1. Birds which crush everything, even the hardest fruits and seeds, in their gizzards. When picking up seeds from the ground, they usually crack and discard their outer coats or shells and destroy the embryos. In the droppings of these birds there was not once found a single seed capable of germination.

2. Birds which leave hard seeds uncrushed, the stones and hard seeds of fleshy fruits passing uninjured through their intestines, while soft seeds and fruits are destroyed. In the droppings of these birds, after feeding, there were found cherry stones as large as 15 mm. in diameter, all of which proved to have retained their germinating power.

3. Birds swallowing small seeds but regurgitating seeds of a diameter of 3–5 mm. or over. A peculiarity of this group of birds is the great rapidity with which seeds pass through the intestines. The seeds of *Sambucus nigra* are evacuated by these birds after only half an hour and those of *Ribes petraeum* after three-quarters of an hour, while other seeds require not more than 1½ to 3 hours. As many as 75–80 per cent of these seeds germinate.

On the basis of these experiments KERNER came to the conclusion that hard seeds and fruits could scarcely be endozoically disseminated by birds, since, as they pass through the bird's gizzard, they are usually ground up. Although this does not hold true in the case of the seeds of fleshy fruits, here we have a different obstacle to their wide dispersal, *viz.*, the short time that they remain in the crop and intestines of the birds that devour them. Hence, KERNER concluded that plants with fleshy fruits may be dispersed by the aid of birds in a single year not more than for several miles and only during a long period of years might gradually be so dispersed over an extensive territory.

As regards the dispersal of seeds by birds carrying them on the surface of their bodies, KERNER considers this quite possible, but he is of the opinion that the number of plants so spread is not great. They embrace, for the most part, shore and swamp plants, particularly, as KERNER established by personal observation, small, annual plants.

MARLOTH ascribes to birds of passage the transport of seeds of northern plants across the equator into South Africa. On the other hand, the Danish ornithologist WINGE (cited by SCHRÖTER, 1934), having investigated thousands of birds of passage that had killed themselves by flying into Danish lighthouses, found that without exception they had empty stomachs and that not a single one had seeds or fruits adhering to their feathers, beaks, or feet. GRÜNBACH and MOSELEY ascribe to the albatross and to many species of birds belonging to the genera *Porcellana* and *Puffinus*, that make their nests in dense vegetation and make annual, long, trans-oceanic flights, an important rôle in the dissemination of seeds. WERTH, on the other hand, reports that on such birds, flying to Kerguelen Island, he did not once find any seeds or fruits adhering to their bodies, and he concludes that, if any had so adhered, they had been washed off in the ocean when the birds alighted to rest.

BIRGER (1907) investigated 170 birds of more than 35 species with the aim of finding seeds or small cakes of earth containing seeds adhering to the surface of their bodies (feet, legs, feathers), but he did not find a single case of such adhesion. As regards the endozoic dispersal of seeds by birds, he found in the stomachs of the birds viable seeds of over 40 species of plants, but he was not inclined to ascribe great significance to this mode of seed dispersal because of the very short time that seeds remain in the digestive tract of a bird. For this conclusion he found further confirmation in the fact that the percentage of species with fleshy fruits in the flora, for instance, of Sweden is very insignificant. If one leaves *Rosa* and *Rubus* out of consideration, species of the latter being concentrated only in southern Sweden, the number of species with fleshy fruits in the flora of Sweden will be only 4.3 per cent. The percentage is somewhat higher, if one includes fleshy-fruited epiphytes growing in Sweden and Norway and also on islands that have recently appeared due to a lowering of the level of the sea. For instance, BIRGER found that on an island, which had existed as such for only four years, out of 112 flowering plants there were only 5 species with baccate or drupaceous fruits, but on an island that had appeared above the level of the water ten years previously, out of 177 flowering plants there were 7 such species and a 22-year-old island out of 191 flowering plants there were 18 species with edible fruits. From these data BIRGER arrives at the conclusion that, in the case of newly stocked territory, species with fleshy fruits form a considerable percentage and this may be ascribed to the rôle of birds in the introduction of seeds.

Unfortunately, however, there has not been reported a single actual observation of the transport by birds of the seeds of any plant that subsequently succeeded in establishing itself and extending its range in the new habitat. All the available data with respect to this as to other kinds of plant dispersal by natural factors have the defect that they cover only separate phases, only separate links in the process of seed dispersal. Thus, we know that some seeds pass through the intestines of birds uninjured, retaining their germinating power; we know, in some cases, the length of time seeds remain in a bird's digestive tract; we know the rate and direction of flight of many birds—yet no one has traced, nor could have traced, the fate of seeds devoured by birds before a flight. Consequently, all the proofs pro and con bear a purely conjectural and very superficial character. The only thing that we can affirm with assurance is that, as regards the seeds of a very few plants, chiefly swamp and littoral plants, it may be presumed that they might be carried by birds to a new locality and take root there.

Significance of Natural Factors in the Distribution of Plants:— Having discussed each of the principal natural factors in plant dispersal, we may now pass to a consideration of the problem as a whole and attempt to find an answer to the question: *Can natural factors in plant dispersal explain all the peculiarities in the present distribution of plants over the earth's surface, or are there other causes that have induced discontinuities in their distribution?*

In the plant as in the animal kingdom, there exists a natural tendency toward multiplication and the winning over of ever new territory for the progeny. To quote DARWIN (*l.c.*, p. 79), "Every organic being naturally increases at so high a rate, that, if not destroyed, the earth would soon be covered by the progeny of a single pair. . . . LINNAEUS has calculated that if an annual plant produced only two seeds—and there is no plant so unproductive as this—and their seedlings next year produced two, and so on, then in twenty years there would be a million plants". Careful calculations made by KERNER have shown that every year, on the average, one plant of *Sisymbrium sophia* produces 730,000 seeds, *Nicotiana Tabacum*—360,000, *Erigeron canadense*—120,000, *Capsella bursa-pastoris*—64,000, *Plantago major*—14,000, *Raphanus raphanistrum*—12,000, and *Hyoscyamus niger*—10,000. Each of these seeds may give rise the next year to a new plant, and each of these plants is capable of producing the indicated number of seeds.

In the lower plants the number of spores formed by one individual plant is far greater. RIDLEY (1930, p. 64) gives the following data on the number of spores produced by a single fungus:—

Psalliota (Agaricus) *campestris*	1,800,000,000
Coprinus comatus	5,000,000,000
Polyporus squamosus	11,000,000,000
Lycoperdon Bovista	7,000,000,000,000

At such a rate of multiplication, even taking into account the small percentage of individuals surviving, if natural factors in plant dispersal really had as much significance as many ascribe to them and the plants dispersed in this fashion could readily become established in the new localities and spread farther, the vegetation of the globe, within the limits of the same climatic zones, would be quite homogeneous.

In addition to the foregoing arguments, we can cite a number of other circumstances that compel us to be very cautious in evaluating the significance of chance factors in the distribution of plants. First of all, it should be stressed that the various adaptations for the transport of fruits and seeds very often do not facilitate the dispersal of plants as much as might be expected. As we noted previously, BENTHAM has shown that those species of *Compositae* having plumed fruits are not, by any means, the most widely distributed species of this family. Similarly, according to the same authority, such adaptations possessed by the fruits of *Compositae* as hooked formations, spines, viscid exudations, etc., to which are usually ascribed very great significance, on close investigation prove not to play such an important rôle. Many of the spiny *Compositae*, such as *Acanthospermum*, have a comparatively limited area. Species of the genus *Echinospermum* (of the family *Boraginaceae*) with prickly fruits are, as a rule, less widely distributed than those of the genus *Myosotis* with smooth fruits. Similarly, the burdock, *Arctium lappa* (of the *Compositae*), characterized by fruits with armed bracts or burrs, has a considerably more restricted range than *Cirsium arvense*.

As another example, we may take CHERMEZON's investigation

(1924) on the dissemination of the seeds of certain sedges. He found that in the fruits of some species of the family *Cyperaceae* there exists a special sponge-like tissue (the cells of which are filled with air) serving as an aid in floating. About thirty species of this family were subjected to study, and these were divided into the three following groups: (1) those with fruits incapable of floating or capable of floating for a very short time; (2) those with fruits capable of floating for a medium length of time; and (3) those with fruits capable of floating for a prolonged period. It was found that the species with fruits entirely incapable of floating have the most extensive areas of distribution. Here, it seems, there must be taken into consideration a number of other factors, such as germination and taking root of seeds, adaptation to new habitat conditions, etc., which it is impossible to gauge precisely.

Moreover, those that are wont to take recourse in chance factors as an explanation of all difficulties encountered in understanding the present-day distribution of plants do not take the trouble to ascertain whether or not there exist the sea currents which, in their opinion, carried the seeds to the necessary place and whether or not they pass in the proper direction for such carriage, or to ascertain the time and routes of flight of birds of passage, whether the beginning of flight coincides with the ripening of the fruits and seeds of the plants under consideration, whether the birds fly at such a time precisely to that place where these seeds must be carried, or whether this period of fruiting coincides with their flight in the opposite direction.

Whenever alleged cases of the responsibility of chance factors of seed dispersal for the distribution of plants have been thoroughly investigated, they have been proved to have no foundation in fact. For example, we may take the critical analysis which Stapf (1894) made of the species in the flora of Mt. Kinabalu (Borneo). The flora of this mountain is composed of species growing likewise not only on other islands of the Indian Ocean as far away as New Guinea but also in the Himalayas. As a result of this analysis, Stapf came to the conclusion that neither birds nor present-day physiographic factors can serve to explain the combination of species growing on Mt. Kinabalu. He found that the higher the altitude, *i.e.*, the more untouched by man the vegetation, the greater the number of endemics, and that there was no correlation whatsoever between the number of endemics and the possibilities for dispersal due to the possession of special adaptations. It would seem that where there was a greater number of plants having such adaptations for dispersal, there should have been fewer endemics, but in many cases the reverse was true.

It was formerly supposed that the lower, spore-bearing plants, whose spores are so easily dispersed by the wind, must have an exceptionally extensive geographical distribution. However, the more their habitats have been subjected to study, the more it has been found that such plants are localized in definite areas, their distribution paralleling that of the flowering plants. Similarly, if the chance transport of seeds over great distances was of such decisive significance, this should be particularly reflected in the distribution of aquatic plants, especially of angiosperms living in the ocean or in large bodies of fresh water.

But even in their case a study of their geography reveals that it is characterized by definite regularities (ASCHERSON, 1905; SETCHELL, 1935). This has been confirmed by a number of papers on the geography of lichens (DEGELIUS, 1935; SCHINDLER, 1936–1938; MINAYEV, 1936, 1938; SUZA, 1925), mosses (HERZOG, 1926), and ferns (CHRIST, 1910).

Barriers to Dispersal: Their Influence on Plant Distribution: — It is necessary to draw particular attention to the existence and significance of *barriers* (physiographic, climatic, edaphic, biotic) to the dispersal of plants, barriers which are often insuperable even in case of the slow, gradual, age-long advance of plants.

1. *Physiographic Barriers.* — Among the barriers hindering a plant's free dispersal over the globe physiographic barriers are of major importance. Chief among these are, for terrestrial plants, *bodies of water*, primarily, of course, seas, and, for aquatic plants, *bodies of land.* KERNER (1898) correctly points out that a considerable number of American plants found their way to Europe not by being carried over by birds of passage nor by the aid of sea or atmospheric currents but simply by man's intervention, and that the boundaries set by the sea are actually insuperable but, at the same time, temporary, *i.e.*, of significance only as long as the present distribution of land and sea is preserved. If at any time a land-bridge had arisen connecting Europe and America, many plants could have migrated over this bridge, and the plants which, as is known, found their way from America to Europe with man's aid might have spread thither without his assistance and become distributed throughout Europe. Environmental conditions would no more have hindered their establishment in Europe than they did after their introduction by man. Precisely the opposite takes place, *i.e.*, migration of land plants is stopped when two regions formerly united (*e.g.*, Spain and North Africa) become separated by the juncture of two water basins (in the given case, the Mediterranean Sea and the Atlantic Ocean).

Another major obstacle in the way of the dispersal of plants we find in *mountains*, the vegetation on the opposite sides of a mountain range often being sharply distinct. Mountains, besides constituting a purely mechanical barrier, hinder the dispersal of plants as a result of the effect they produce on climatic conditions. Even in case seeds are accidentally carried from one side of a mountain range to the other, the differences in climate create for the immigrant plants habitat conditions to which they have not been accustomed.

2. *Climatic Barriers.* — Climatic conditions constitute one of the chief factors in the distribution of plants over the earth's surface, determining in most cases the limits of the distribution of species. Due to the close dependence of plants on climatic conditions, climatic and vegetation zones closely correspond. The ability to become adapted to entirely different habitat conditions is exceedingly limited, in consequence of which climatic boundaries constitute very real barriers to plant dispersal. It should be noted that in this regard not only is the climate as a whole of significance but each separate climatic factor. In some cases the major rôle is played by temperature conditions; in others by the length of day (short- and long-day plants); in still

others by the humidity of the air (some plants cannot withstand a dry atmosphere and perish even when there is adequate soil moisture; others cannot endure a humid atmosphere). Hence, a plant accidentally brought into a region characterized by climatic conditions unsuited to it will be unable to develop and will be doomed to perish, even if its seeds germinate.

3. *Edaphic Barriers.* — What we have just said about climatic conditions holds true for other habitat conditions, particularly edaphic conditions, such as the physical and chemical structure of the soil, soil humidity and temperature, etc., which in combination or separately may constitute a barrier to the spread of a species or limit its range to only those localities characterized by certain definite conditions.

The foregoing, however, are not all the difficulties that a species must overcome in its conquest of new territories. Synecology, or the science of the interrelations of plants in plant communities, has shown us what great significance such interrelations have in the distribution of plants. As BRAUN-BLANQUET (1928) has emphasized, the chief difference between plant and animal communities is that the former are characterized by a struggle for existence not limited by any principles of usefulness, division of labor, communal toil, or by any other factors arising from the conscious activity of organisms forming part of a biocoenosis.

This struggle for existence, or competition between separate species and individual plants, is comparatively insignificant in so-called "open associations", on sands, cliffs, and solonchak soils, where the vegetation does not form a continuous carpet but is spotted, with larger or smaller unoccupied spaces between the plants. But here habitat conditions are usually so specific and unfavorable that only a few species are adapted to them. In other localities, where edaphic and physiographic conditions are more favorable, the vegetation forms "closed associations", the members of which utilize every bit of soil, every ray of sun. In such associations competition between the plants, with respect to space, light, and food, constitutes a very characteristic feature. There is no call for us here to enter into a discussion of the problem of the struggle for existence in the plant world. If we have touched upon this problem, it has been only with the aim of pointing out that even among the members of a given association, as a result of natural selection and the struggle for existence, there survive only an insignificant percentage of the progeny, and that, consequently, the chances of a new, would-be member that happened to spring up there from an accidentally introduced seed to gain full citizenship rights in such a closed association are almost nil.

The foregoing facts make it necessary to regard quite skeptically the statements of old authors who but slightly delved into these problems in plant life (phytosociology) and also those of modern authors who exaggerate the significance of chance factors in the distribution of plants over the globe. We believe that we have made it sufficiently clear that it is practically impossible for an accidentally introduced seed to give life to a plant that can enter into the composition of an alien association. In order for a plant to become established in an entirely new region, special conditions are necessary, *viz.*, territory suitable for plant life but unoccupied or not fully occupied by other plants.

Stocking of Unoccupied Territories: — Territories unoccupied by vegetation, if we exclude those places where plant life is impossible, no longer exist on the globe. The entire surface of the earth, every spot where conditions are at all fit for the existence of organisms, is inhabited. But unoccupied areas may arise as a result of chance, catastrophic circumstances—alluvial deposits destroying the vegetation buried beneath them, avalanches, volcanic eruptions covering huge areas with ash and lava, upheaval of land from beneath the surface of the sea—or as a result of the destruction of vegetation by man.

According to COMES, the first plants that made their appearance on the Vesuvian lava fields after they had become cooled were unicellular *Algae*, next came lichens, then mosses and ferns, and, lastly, the flowering plants. Hence, it appears that for the higher, flowering plants to become established it is necessary that sterile soils should first be subjected to certain changes brought about by the activities of lower organisms (CHEVALIER and GUÉNOT, 1925). ULBRICH (1928) watched for a number of years the invasion by plants of a former gravel pit near Berlin. This pit was abandoned in 1894 and during the subsequent years was gradually invaded by plants. In 1900 there were found on its territory 109 species, including 50 cryptogams; by 1903 there were already 268 flowering plants and 113 cryptogams; by 1922—a total of 959 species, of which 429 were flowering plants. Most of the latter were plants whose seeds and fruits are wind-dispersed. A large number of the plants belonged to species that grew in the surrounding territory, but there were also many that grew only at a distance of 150 km. from the place of observation.

Very often, when a territory is deprived of its plant covering by some catastrophe, there at first appears on it a very homogeneous vegetation, sometimes arising as a result of the rapid spread of a single species, which subsequently begins to die out as a result of competition between its own individual plants and then, little by little, is crowded out by the later comers that gradually put in their appearance. An interesting case of such a rapid spread of a single species is reported by MOLISCH (1927), who observed it immediately after the 1923 earthquake in Japan. In Tokyo the trunks of the trees along the sides of the street and in the parks, due to the widespread fires throughout the city, were badly charred, and, consequently, all the epiphytic vegetation on them was destroyed. This territory, free of competitors, was very rapidly inhabited by the fungus, *Monilia sitophila*, which soon covered all the charred tree trunks. Ordinarily this fungus is rarely encountered in Japan, MOLISCH having found it only infrequently on discarded ears of corn lying on damp ground.

An interesting example of the stocking of unoccupied territory, not far distant from occupied land, is provided by the Åland Islands. The entire process of the colonization of these islands by plants was subjected to detailed study by PALMGREN (1922, 1925, 1929), who found that such a process depends on the size of the territory being colonized and the biological peculiarities of the immigrant species. The upheaval of these islands began in the post-glacial period and is still in progress, in the course of each century the surface of the islands rising higher by about 0.5–0.6 m. Consequently, we have here, as a result of this secular land-rise, a rare opportunity to study the relatively recent stocking of virgin territory. Moreover, the study of the history of the vegetation of these islands is of added interest, since the gradual uplifting of territories from the sea cannot but have significance as regards the composition and character of the flora, which undoubtedly would have been quite different had the upheaval of the islands taken place all at once. These circumstances create a wide field for the study of the gradual development of a plant covering and the distribution of species.

As a result of his investigations, PALMGREN found that there existed a correlation between the number of species and the size of the territory occupied. In the region studied by him he found three categories of territories: (1) the smallest, characterized by 153–164 species; (2) territories 3–4 times as large as those in the first category and having 202–203 species; and (3) territories twice as large as those in the second category and having 210–216 species. Hence, it follows that a definite territory, provided certain conditions hold true, is characterized by a definite number of species, or, in other words, that the larger the territory, the greater the number of species. What is the cause of this correlation between the size of a territory and the number of species inhabiting it? PALMGREN considers it to be simply the absence of free space for a greater number of species. Any given territory may embrace only a definite, limited number of species. Plants invading a territory already occupied by the maximum number of species that it can maintain are confronted by a closed association barring entrance to new elements.

What determines precisely one and not another combination of species colonizing a given territory? PALMGREN (1929) considers the most important factor to be *chance*. The species that reach a given territory first, due either to certain adaptations they possess or to some other reasons, have an advantage over late comers in gaining a foothold in that territory. Hence, though the number of species in a given territory is more or less fixed, which particular species will inhabit it is indefinite, depending on which species chance to invade the territory first. These results arrived at by PALMGREN have been confirmed by VALOVIRTA (1937), who conducted similar investigations on the Quarken Islands in the Gulf of Bothnia.

The regularities established by these investigators with respect to the limited number of species that may occupy a definite territory do not arise, however, solely by chance, since they must necessarily depend also on the ecological conditions in the territory being colonized, as EKLUND (1931) has pointed out in his criticism of PALMGREN's conclusions. Nevertheless, the data presented by PALMGREN are of interest, since they make clear the difficulties that a species dispersed by natural factors must, under present-day conditions, overcome, in case it encounters a plant community already possessing its full quota of species. They provide yet another proof of the insignificance of natural factors in the dispersal of plants over great distances, except in those cases when the plants chance to come upon unoccupied territory that for some reason or other has not yet been colonized.

For the study of the resurrection of vegetation destroyed as a result of a volcanic eruption Krakatau has been considered a particularly suitable subject. Observations on the revegetation of this island have already been made for over 50 years, as a result of which there have been accumulated many scientific papers devoted to it. In recent years there have appeared two extensive works (ERNST, 1934; DOCTERS VAN LEEUWEN, 1936), reviewing the results of these investigations. Krakatau, one of the Sunda Isles in the Malay Archipelago, is a volcanic island, approximately 14 sq. mi. in area and located 25 mi. from Java and 23 mi. from Sumatra. In 1883, as a result of a big volcanic eruption, the luxurious vegetation of the island was destroyed by a downpour of hot ashes and stones that covered the entire island in a thick layer. Regarding the former vegetation of the island we have no exact data. It is only known that two weeks prior to the end of the eruption, at a time when in the northern part of the island there could be seen only occasional bare trunks of trees, the whole southern slope comprising the more elevated part of the island was still covered, according to observations of the geologist VERBEEK, by dense, green vegetation, and only during the very last moments of the eruption was this part of the island also buried under ashes and pumice.

Three years after this catastrophe TREUB (1888) made a trip to the island for the purpose of studying the new vegetation on it. On the shores of the island he found about 16 different species of flowering plants, and in the interior 8 species of flowering plants (6 of which were not found on the shores) and 11 species of ferns, the latter in abundance. Assuming that all the original flora of the island, including seeds and underground parts of plants, had been completely destroyed at the time of the eruption and that, due to the uninhabitableness of the island, plants could not have been brought by man, TREUB came to the conclusion that sea currents and wind, and later also birds, constituted the factors responsible for the stocking of this island with plants. From that time on, TREUB's work was regarded as an example *par excellence* of the significance of natural factors in the dispersal of plants, as proof based on the actual observation of facts, and TREUB's final conclusion almost as a law. Since then, from time to time, various investigators have visited Krakatau and published their findings as regards its vegetation. The number of plants collected constantly increased, but the mode of their origin on the island did not evoke any doubts. For most of these investigators TREUB's premise as to the complete destruction of the vegetation on the island was regarded as an incontrovertible fact.

In 1929 BACKER, who had himself visited Krakatau on numerous occasions and who held the same viewpoint as all the other botanists, undertook a summary of the data of all the investigators of the island's vegetation, a survey of all the results of the various expeditions to the island, including those in which he had personally participated. Having completed this survey, he was forced to renounce his previous views and to put the whole question in an entirely different light. He found that it was by no means established that *all* the vegetation of Krakatau had been destroyed by the volcanic eruption of 1883, and he concluded, therefore, that the problem of Krakatau can never be solved and should be regarded as without significance for botanical science.

It was, of course, a major error on the part of the first investigators of the flora of Krakatau not to give adequate attention to such an important circumstance as the possibility of seeds, roots, or stems having been preserved in the soil. Although both ERNST and DOCTERS VAN LEEUWEN consider that such parts of plants, due to the excessively high temperature of the soil buried under the burning ashes of the volcano, could not have been preserved, doubts as to their complete destruction can always arise, as long as they were not eliminated at the very first by special investigation. This is all the more true, since we now have data of the preservation of seeds and the underground parts of plants after eruptions of other volcanoes, *e.g.*, after that of the volcano Kamagatake, in southern Hokkaido (Japan). According to JOSHI (1932), at the time of the eruption of this volcano in 1929, which lasted for two days, an area of 5,000 hectares on its slopes and at its foot was buried under a layer of ashes and pumice, averaging one meter in thickness and in some places as much as three meters thick. The impenetrable forest that had previously covered its slopes was completely burned and destroyed by the products of the eruption. JOSHI for two years closely watched this area, and he found that there had remained in the soil viable, though injured, underground parts of plants that rapidly produced new shoots. Due to the irregularity of the relief, the depth to which the soil was covered by ashes and pumice varied and in some places was very inconsiderable. In such places as early as two months after the eruption revegetation commenced, and after a year had elapsed there appeared perennials that had developed from underground parts that had survived.

Nevertheless, there is no doubt that BACKER's conclusion that a study of the revegetation of Krakatau altogether lacks scientific interest must be regarded as extreme. Even if part of the vegetation was restored as a result of the preservation of seeds or underground parts of plants, still a goodly proportion of the 324 species making up, according to DOCTERS VAN LEEUWEN, the flora of Krakatau at the present time were brought to the island by natural factors of dispersal. Krakatau constitutes an interesting case of the stocking of unoccupied territory as a result of the transport of seeds over a relatively small distance (ca. 25 miles) by sea currents, wind, and birds.

Conclusions: — *Plants are dispersed as result of the transport of their seeds by natural means only slowly and gradually, gaining new territory step by step.* The transport of seeds by such means over great distances may be a factor in plant dispersal only in case the seeds encounter a territory where ecological conditions are favorable for the given species and where the original plant associations have been destroyed (which otherwise would constitute an insuperable obstacle to the newcomer's becoming established) or where there exist open formations, as, for instance, on the seashore. Consequently, the transport of seeds by wind, sea currents, birds, or other natural means cannot serve as an explanation of discontinuous areas with widely separated parts. Numerous facts, *e.g.*, the occurrence of endemic species isolated from one another on separate peaks of the same mountain chain, point to an entirely different mode of origin of such areas. Hence, we must seek an explanation of those moments in the geography of plants that are incomprehensible in the light of present-day factors *not* in the action of chance circumstances but in the connection existing between plant distribution and the past history of our globe.

References:

Amann, I., 1894: Woher stammen die Laubmoose der erratischen Blöcke der schweizerischen Hochebene und des Jura? (Ber. d. Schweiz. Bot. Ges., Vol. 4).

Ascherson, P., 1905: Die geographische Verbreitung der Seegräser, 3rd ed. (Anleitung z. wissensch. Beobachtungen auf Reisen, herausgeg. von Neumayer, Hannover).

Backer, C. A., 1929: The Problem of Krakatau as Seen by a Botanist (Buitenzorg).

Beccari, O., 1916: Il Genere *Cocos* Linn. e le Palme affini (Florence).

Béguinot, A., 1912: Osservazioni e Documenti sulla Disseminazione a Distanza (Padua).

Bentham, G., 1873: Notes on the classification, history and geographical distribution of *Compositae* (Journ. Linn. Soc., Botany, Vol. 13).

Birger, S., 1907: Über endozoische Samenverbreitung durch Vögel (Svensk Bot. Tidskrift, Vol. 1, No. 1).

Braun-Blanquet, J., 1928: Pflanzensoziologie (Berlin).

Brockmann-Jerosch, H., 1916: Die Anschauungen über Pflanzenausbreitung (Vierteljahrsschr. Naturf. Ges. Zürich, Vol. 61).

Brockmann-Jerosch, H. and M., 1923: Betrachtungen über Pflanzenausbreitung (Verh. d. Naturf. Ges. in Basel, Vol. 35).

Chermezon, H., 1924: Sur la dissémination de quelques Cypéracées (Bull. Soc. Bot. d. France, Vol. 71).

Chevalier, A. et Guénot, L., 1925: Biogéographie, Chap. X. Peuplement des espaces vides (Paris).

Christ, H., 1882: Pflanzenleben der Schweiz (Zürich).

Christ, H., 1910: Die Geographie der Farne (Jena).

Cook, O. F., 1901: The origin and distribution of the coco palm (Contrib. U.S. Nat. Herb., Vol. 7, No. 2).

Degelius, G., 1935: Das ozeanische Element der Strauch- und Laubflechten-Flora von Skandinavien (Acta Phytogeogr. Suec., Vol. 7).

Docters van Leeuwen, W. M., 1936: Krakatau, 1883–1933 (Ann. Jard. Bot. Buitenz., Vols. 46–47, pp. 1–506) (see Bibliography).

Eklund, O., 1931: Über die Verbreitung der Schärenflora Südwest Finnlands (Acta Bot. Fenn., Vol. 8).

Ernst, A., 1934: Das biologische Krakatauproblem (Vierteljahrsschr. Naturf. Ges. Zürich, Vol. 89).

Falck, R., 1927: Über die Grössen, Fallgeschwindigkeit und Schwebewerte der Pilzsporen und ihre Gruppierung mit Bezug auf die zu ihrer Verbreitung nötigen Temperaturströmungs-Geschwindigkeiten (Ber. d. deutsch. Bot. Ges., Vol. 45, No. 5).

Guppy, H. B., 1917: Plants, Seeds, and Currents in the West Indies and Azores (London).

Heintze, A., 1932–36: Handbuch der Verbreitungsökologie der Pflanzen, Parts I–III (Stockholm).

Hemsley, W. B., 1885: Report on the Scientific Results of the Voyage of H.M.S. Challenger, Vol. I, Botany (London, Edinburgh and Dublin).

Herzog, Th., 1926: Geographie der Moose (Jena).

Hooker, W., 1827: *Lodoicea sechellarum* (Curtis Bot. Mag., N.S. Vol. 1, tab. 2734–2738).

Ilyinsky, A. P., 1933: Area and its dynamics (In Russian; Soviet Bot., No. 5).

Joshi, J., 1932: Revegetation of the volcano Kamagatake after the eruption in 1929 (Bot. Mag. Tokyo, Vol. 46).

Keller, B. A., 1922: On fruits and seeds distributed by spring floods over inundated river basins (In Russian; Russian Hydrobiol. Jour., Vol. 1).

Kerner, A., 1871: Der Einfluss der Winde auf die Verbreitung der Samen im Hochgebirge (Ztschr. d. deutsch.-oesterr. Alpenver.).

Kerner, A., 1879: Beiträge zur Geschichte der Pflanzenwanderungen (Oest. Bot. Zeitschr., Nos. 6, 7).

Kerner, A., 1898: Pflanzenleben, Vol. II, 2nd ed. (Leipzig).

Kotilainen, M., 1929: Über das boreale Laubmooselement in Ladoga-Karelien (Ann. Soc. Zool.-Bot. Fennicae, Vol. 11, No. 1).

Merrill, E. D., 1926: Enumeration of Philippine Plants, Vol. IV (Manila).

Minayev, N. A., 1936: New lichens in the flora of the vicinity of Leningrad (In Russian; Acta Inst. Bot. Acad. Sci. U.R.S.S., Ser. II, No. 3).

Minayev, N. A., 1938: Relic elements in the lichen flora of the eastern part of the Baltic States (*In* "The Problem of Relics in the Flora of the U.S.S.R."; in Russian; Leningrad).

MOLINIER, R. et MÜLLER, P., 1938: La dissémination des espèces végétales (Rev. Gén. Bot., Vol. 5, Nos. 590, 592, and ff.).

MOLISCH, H., 1927: Im Lande der aufgehenden Sonne (Vienna).

MÜLLER, P., 1933: Verbreitungsbiologie der Garigueflora (Beih. Bot. Centralbl., Abt. II, Vol. 50).

MÜLLER, P., 1934: Beitrag zur Keimverbreitungsbiologie der Endozoochoren (Ber. Schweiz. Bot. Ges., Vol. 43, No. 2).

MÜLLER, P., 1935: Über Samenverbreitung durch den Regen (Stat. Géobot. Méditerran., Communic. No. 43).

MURBECK, S., 1919, 1921: Beiträge zur Biologie der Wüstenpflanzen (Lunds Univ. Årsskrift, N. F. Vol. 15, No. 10, and Vol. 17, No. 1).

PALMGREN, A., 1922: Über Artenzahl und Areal, sowie über die Konstitution der Vegetation (Acta Forestalia Fennica, Vol. 22).

PALMGREN, A., 1925: Die Artenzahl als pflanzengeographischer Charakter, sowie der Zufall und die sekuläre Landhebung als pflanzengeographische Faktoren (Fennia, Vol. 46, No. 2).

PALMGREN, A., 1929: Chance as an element in plant geography (Proc. Intern. Cong. Plant Sciences, Ithaca 1926, Vol. 1, pp. 591–602).

PETTERSSON, B., 1929: Ferns and flowering plants on erratic blocks with special reference to their modes of dispersal (Memoranda Soc. pro Fauna et Flora Fenn., Vol. 6, pp. 25–47).

PRAEGER, LLOYD, 1923: Dispersal and distribution (Jour. Ecol., Vol. 11).

RIDLEY, H. N., 1905: On the dispersal of seeds by wind (Ann. Bot., Vol. 19).

RIDLEY, H. N., 1930: The Dispersal of Plants throughout the World (Ashford).

SCHARF, R. F., 1925: Sur le problème de l'île de Krakatau (C. R. Congr. Ass. Franç. de Grenoble) (Biogéogr.).

SCHIMPER, A., 1891: Die Indo-malayische Strandflora (Mitteil. aus d. Tropen, No. 3).

SCHINDLER, H., 1936, 1937, 1938: Beiträge zur Geographie der Flechten (Ber. d. deutsch. Bot. Ges., Vol. 54, No. 10; Vol. 55, Nos. 3 and 9; Vol. 56, Nos. 1, 8, and ff.).

SCHMIDT, W., 1918: Die Verbreitung von Samen und Blütenstaub durch die Luftbewegung (Oester. Botan. Zeitschr., Vol. 67).

SCHRÖTER, C., 1934: Genetische Pflanzengeographie (Handwörterb. Naturwiss., 2nd ed., Vol. IV).

SERNANDER, R., 1927: Zur Morphologie und Biologie der Diasporen (Acta Reg. Soc. Sci. Uppsala).

SETCHELL, W. A., 1935: Geographic elements of the marine flora of the North Pacific Ocean (Amer. Naturalist, Vol. 69, No. 725).

SOLMS-LAUBACH, H., 1905: Die leitenden Gesichtspunkte einer allgem. Pflanzengeographie, V. Besiedelung des Standorts durch die Art (Leipzig).

STAKMAN, E. C., HENRY, A. W., *et al.*, 1923: Spores in the upper air (Jour. Agr. Res., Vol. 24, No. 7).

STAPF, O., 1894: On the flora of Mount Kinabalu (Trans. Linn. Soc., Ser. 2, Part IV, Botany, p. 110).

SUZA, J., 1925: A sketch of the distribution of lichens in Moravia with regard to the conditions in Europe (Publ. Fac. Sci. Univ. Masaryk, No. 55).

TREUB, M., 1888: Notice sur la nouvelle flore de Krakatau (Ann. Jard. Bot. de Buitenzorg, Vol. 7).

ULBRICH, E., 1928: Biologie der Früchte und Samen (Berlin).

VALOVIRTA, E. J., 1937: Untersuchungen über die sekuläre Landhebung als Pflanzengeographischer Factor (Acta Bot. Fenn., No. 20).

VOGLER, P., 1901: Über die Verbreitungsmittel der schweizerischen Alpenpflanzen (Flora, Vol. 89, Ergänzungsband).

ZOHARY, M., 1937: Die Verbreitungsökologischen Verhältnisse der Pflanzen Palästinas (Beih. Bot. Centralbl., Vol. 56, Abt. A, No. 1).

Chapter IX

THE MIGRATIONS OF SPECIES AND FLORAS AND THEIR CAUSES

As we have shown in the preceding chapter, plant dispersal in most cases is not "saltatory", *i.e.*, it is not achieved as a result of the chance transport of seeds over *great* distances. Plants usually extend their range slowly, gradually, step by step, as a result of the dissemination of their seeds by natural factors over territories located *near* the mother plants and not extending far beyond the boundaries of the latter's area. Such slow extensions of area may, moreover, affect not only separate forms or varieties but also species and even whole plant communities. The fact that the progeny of any plant—by means of various autochoric processes and such natural agencies as wind, water, and animals—usually takes root at *some* distance (but not a *great* distance) from the mother plant results in the slow, gradual advance of a species, in an expansion of its area, continuing until the given species encounters some obstacle—physiographic, climatic, edaphic, or biotic—hindering its further spread. Changes within the plants themselves occurring, as a result of natural selection, during the process of their dispersal often enable them to overcome some of these obstacles and so to continue their spread. Nevertheless, such extensions of area lack any definite direction and are inadequate to explain those moments in the history of our globe when entire floras were in a state of motion, changing their areas of distribution in one and the same direction. The question arises: What are the causes underlying such mass movements of floras and their component species?

Migration Theory: — The distribution of plants and their grouping in floras are primarily controlled, as we know, by *climatic conditions*, which determine the geographical zones and altitudinal belts in which their areas are located. Within these climatic limits plant distribution is further controlled by edaphic conditions. The latter are of secondary significance, being indefinite in character and not subject to any regular zonation as are climatic conditions. If, however, climatic zonation were unalterable, as was until recently presumed, the distribution of plants over the globe would be fixed and the species composition of floras would not be characterized by such diversity as exists. Paleobotanic data and analyses of the areas of species in present-day floras provide clear evidence that time and again during the history of our earth—in all geological epochs down to the last post-glacial period—there have occurred great movements of floras. That such movements have actually occurred is likewise testified to by the fact that plant fossils differ in their geographical and altitudinal distribution from the living representatives of the same genera and by the mixed character of the species composition of floras.

In view of the fact that the dispersal of species takes place as the result of the spread of individual plants during the process of their multiplication and continues until the climatic boundaries are reached, provided that no other obstacle has previously barred their advance, we must arrive at the only conclusion possible, *viz.*, that successions of floras result primarily from changes in the location of the climatic zones. It may thus be considered as definitely established that climatic zones have always existed on the earth but that the location of these zones has changed many times. Furthermore, the upheaval of whole mountain systems and the drying up of seas have created new territories, free from competition, whither species have drifted in their migrations. Spreading from slopes to mountain peaks and from forest communities to the arid beds of former seas, these migrants underwent processes of form-genesis, whole cycles of polymorphic forms and geographical series of species often being created.

Hence, summarizing what has been said, we may presume that the present distribution of plants is the result not only of the autochthonous development of floras but also of the migrations of species and floras (due to climatic and physiographic changes), migrations made possible by the dissemination of seeds and modified by the character of the topography and the ecological conditions of the given territory. As Blytt (1882) so aptly stated, present-day vegetation reflects, as in a mirror, the geological history of a country, different groupings of species being the expression of different stages in that history.

In this concept we fail to find, however, an explanation of why and how, as a result of climatic changes, the areas of species are altered. "We cannot say," wrote Darwin ("Origin of Species," 6th ed., 1911, p. 533), "why certain species and not others have migrated; why certain species have been modified and have given rise to new forms, whilst others have remained unaltered. . . . why one species ranges twice or thrice as far, and is twice or thrice as common, as another species within their own homes".

It is, nevertheless, possible to elaborate a working hypothesis capable of explaining how successions of floras may have taken place. Such a hypothesis has been postulated by Good (1931) under the name of the "theory of tolerance". This theory implies "that a species is able to occupy only those parts of the world where the external conditions are within those of its range of tolerance. This total area which a species can occupy in virtue of its tolerance is conveniently termed its 'potential area'. The size of the potential area will tend to vary with change in external conditions" (p. 155).

If species possessed unlimited tolerance, unlimited capacity for adaptation and for the formation of new races adapted to any changes that might occur in habitat conditions, and if these adaptations occurred with the same rapidity as the climatic changes, then species would be able to remain within the limits of their original areas. In a few cases this has actually taken place; relic species and areas testify to this. In various floras we may find species that have, by adapting themselves to new habitat conditions, undoubtedly survived from former geological periods characterized by different climatic conditions. This holds particularly true in those regions where a cold climate has been replaced by a warmer one, as, for instance, in South Africa.

As a rule, however, species do not possess unlimited tolerance nor the capacity to adapt themselves to changing conditions as rapidly as these changes occur. This was pointed out by DARWIN (*l.c.*, p. 421), who remarked that in nature species "probably change much more slowly, and within the same country only a few change at the same time. This slowness follows from all the inhabitants of the same country being already so well adapted to each other, that new places in the polity of nature do not occur until after long intervals, due to the occurrence of physical changes of some kind, or through the immigration of new forms. Moreover variations or individual differences of the right nature, by which some of the inhabitants might be better fitted to their new places under the altered circumstances, would not always occur at once".

Changes in generic characters proceed still more slowly. Imprints (buried under the earth for hundreds of thousands of years) of many fossil angiosperms of the Tertiary period, which to this day form part of several present-day floras, closely resemble modern representatives of the same genera. As DARWIN (1868, Chap. 26, p. 352) put it, "Generic characters are less variable than specific characters; and the latter are those which have been modified by variation and natural selection, since the period when all the species belonging to the same genus branched off from a common progenitor, whilst generic characters are those which have remained unaltered from a much more remote epoch, and accordingly are now less variable".

It is well known that as a result of changes in habitat conditions there may arise, by mutation, new races, enabling a species to expand its area beyond the limits restricting the spread of the initial species, but there is no case known of such mutational changes embracing entire floras. They affect only separate species and are more or less accidental in character. These considerations give grounds for presuming that, as a rule, the evolution of the biological and physiological characteristics of plants, involving adaptation to new habitat conditions, cannot proceed as rapidly as changes in habitat conditions. Consequently, in case of such changes, a plant must change its place of abode or perish. It is able to change its place of abode, thanks to various adaptations for the dissemination of its progeny. The possession by a plant of the capacity to tolerate within certain limits changes in habitat conditions serves the plant, when it finds itself outside its "range of tolerance", as an incentive to change its habitat in the direction of a return to environmental conditions suitable for it, *i.e.*, as an incentive to migration. Let us quote GOOD (*l.c.*, pp. 157–9): —

"Taking into account the great number of plants species and the comparatively narrow world gamut of many external conditions it is not difficult to imagine that more than one species may have similar ranges of tolerance, especially as regards some factors. This is strictly in accordance with the observed facts of plant competition and is a satisfactory explanation of it. It supposes that the tolerances of the two species, although similar in the main, are actually very slightly different and that this difference is extremely important in deciding the result of competition. This result may be a balanced deadlock, suggesting that there is nothing to choose, in tolerance, between the species. More often competition will result in the establishment or development of one species at the expense of the other, and this is assumed to mean that, in some very subtle way, the tolerance of the victorious species is more closely correlated with the actual conditions. . . .

"Since the tolerances of closely related species need not be closely comparable, it follows that the generic tolerance as regards single factors may be continuous or discontinuous. If the former, the generic tolerance will include a complete range of values for a particular factor, from generic minimum to generic maximum. If the latter, there will be gaps in the range, these gaps representing factor values not included in any specific tolerance."

In order to show just how the tolerance of a species is able, in case of changes in external conditions, to induce its migration for the purpose of finding suitable habitat conditions, GOOD takes as an example a simple and purely hypothetical case of a species whose tolerance is determined by one external factor only, temperature (actually, of course, several factors are always involved). GOOD (p. 161) presents this example as follows:—

"Let us imagine a species having a mean annual tolerance of 10°–20° C. . . . and let us suppose that the area over which these temperature values prevail, the potential area of the species, is continuous. Let us further imagine that the species has covered the whole of this area so that the area of the species and that of the appropriate temperature conditions are coterminous.

"Every generation, if the species is herbaceous and monocarpic, and every reproductive season if the species is perennial, the individuals comprising it will produce dispersal units (usually seeds or fruits) and these will tend to be disseminated in all directions from the parent plants".

In case these seeds germinate within the existing area of the species, the plants arising therefrom will have all the necessary conditions for normal growth and development. If, however, these seeds are carried beyond the boundaries of the area of the species, *i.e.*, beyond the range of tolerance of the species to temperature conditions, then, if there do not arise mutant forms with a different range of tolerance, the plants produced from these seeds will perish or, at any rate, will not develop normally. "This process", to quote GOOD (p. 162), "will continue at reproductive intervals of time as long as the specific tolerance and distribution of temperature remain unchanged, a proportion of the units at each reproduction failing to develop".

"Now suppose", continues GOOD, "that a climatic change associated with general lowering of temperature begins. Other things (such as topography) being equal, the temperature area of 10°–20° C. will move towards the equator. What will be the effect upon the plants? On the northern edge of their area, as climatic movement begins, there will be the equivalent of a contraction of potential area so that some of the dispersal units, not only of the outermost plants but also of those slightly further in, will fall outside the necessary conditions. Before very long the parents which were originally the outermost will be themselves outside the potential area and will therefore perish. As the climatic movement continues the belt of destruction in its wake will widen.

"On the southern edge of the specific area the conditions will be exactly reversed. After a time none of the dispersal units of the outermost individuals will any longer fall outside the potential area, but within it, and will mature successfully, producing dispersal units in their turn. These new individuals will at first disseminate themselves partly outside the area, but very soon, with the continuance of climatic change, their dispersal units, too, will fall within the necessary conditions, and this process will be repeated in succeeding generations [until, of course, the climatic movement ceases].

"The combined effect on the southern and northern edges of distribution, together with similar but modified effects on the flanks, will in fact be such that correlation is maintained between climatic and specific area, and hence, since the former moves, so also does the latter. *But this movement will only result if tolerance remains unchanged while climate alters*". (Italics in original).

This, as GOOD himself states, is "the most crucial point of the whole theory", for, if both tolerance and climate changed at an equal rate, there would be no plant movement and the entire theory would collapse. But usually this does not occur, since evolutionary changes in a species, as we pointed out above, proceed considerably more slowly than changes in habitat conditions. The movements of floras known to us suffice to show that these changes are not simultaneous. They may coincide in time and rate only in case there occurs a mutation affecting the tolerance of the species.

Quoting GOOD further (pp. 163–165): —

"In the simple case described above there is a very great difference between the individuals in the van of the movement and those in the rear. The correlation between climate and area in the van is never seriously upset: there is simply a gradually unfolding area into which dispersal will be effective. The conditions in the rear are quite different. Here the potential area is continually diminishing and the possibility of successful dispersal is, for many individuals, becoming increasingly small, so that the plants are constantly in incomplete harmony with their external conditions. . . .

"This hypothetical example of the working of tolerance is, as has been stated above, the simplest imaginable and in nature the circumstances will almost always be more complex. . . .

"In the example given, only one species is supposed to be affected by climatic change, but this is clearly an ideal condition. In all normal circumstances, climatic or other external change will affect a number of species, generally of varied types. It is, moreover, highly improbable that the tolerances of all the species will be alike. They will most likely vary greatly. The realisation of this leads to a most important conception in the theory: its ability to exert a selective or sifting effect among species.

"Suppose, as another purely hypothetical illustration, that the climatic change of temperature already taken as an example affects one hundred species instead of only one. Allowing for an average amount of variation in specific tolerance it is likely that the change will react upon different species in different ways. Some it may affect directly and immediately, but others it may affect either indirectly or not at all, *their* ranges being determined more particularly by their relations to factors other than temperature. This being so, change of temperature will neither reduce their potential areas in one direction nor increase it in others and they will remain unmoved. The total result of this will be that the temperature change will cause movement in some species but not in others. In terms of movement, some will advance in the direction of change and others will remain stationary so that there is a selective effect. . . . .

"In other directions, too, the original simple scheme must be augmented if it is to give anything like a complete picture of the effects of the theory. For example a climatic change will rarely be confined to a limited region. A general lessening of temperature, like that imagined, would in all probability extend over at least a large part of a hemisphere and even perhaps over the whole world, with the result that there would be a general movement towards the equator of parallel temperature zones. In accordance with the postulated effect of tolerance, this means that one geographical zone will sooner or later come to possess the flora originally characteristic of an adjoining zone. The area rendered unsuitable for one set of species will become in turn the potential area of another set. This, combined with the sifting effect just described, gives a picture, not only of floral migration in bulk, but also of the intermingling of diverse floral elements. Consider the floras of two adjacent parallel climatic regions affected by, let us say, temperature change. Since presumably only a portion of the species in each zone (the proportion may be very high) is susceptible to this particular external change, some species will move under its influence and others will not. But a similar selection is at work in both zones and the result will be that the species left behind by the movement of the zone in advance will become mixed with the species which have advanced with the movement of the zone in rear".

Moreover, it should be kept in mind that a climatic change, usually involving changes not in one but in several factors, induces changes in all the habitat conditions (*i.e.*, in all the ecological conditions affecting the habitat or habitats) of a species. This was pointed out by Kashkarov and Korovin (1931), who showed, on the basis of the migration of different groups of vegetation in Soviet Central Asia, that there exists a correlation between plant migration and changes in habitat conditions. Thus, an alteration in ecological conditions, *e.g.*, the expansion of desert areas, creates new routes of dispersal (*viae oecologicae*, using these writers' term) for sand vegetation. These new routes may embrace extensive territories, if habitat conditions are homogeneous, as was the case, for example, in the Tertiary period in Eurasia. A change in ecological conditions, accompanied by an advance of these routes of dispersal into new territories, constitutes a factor in the further modification and dispersal of both plants and animals.

As a result of the migrations of species, both the size and shape of their areas are altered, since it is impossible to imagine that such migrations could occur in so regular a fashion that the areas could retain their original contour. The diverse topography of our globe, the distribution of land and sea, excludes the possibility of an area retaining its former shape. Consequently, as a result of their migrations, the areas of some species increase in size, as others decrease. A continuous area may, due to topographical conditions, break up into two or more parts, becoming a discontinuous or even insular area. Furthermore, the potential area of a species may increase in size at a considerably more rapid rate than the actual dispersal of the species, the result being that the species, temporarily or permanently, ceases to occupy the whole of its potential area.

The processes involved in the geographical distribution of plants require for their realization exceptionally long periods of time, far exceeding the period of man's life upon this planet. Hence, there can be no thought of direct observation or experimental proofs of the foregoing postulates; only by indirect proofs can their soundness be established. As one of such indirect proofs we may cite the fact that the individuals of any one species react in a more or less similar manner to external conditions, resembling one another in tolerance as well as in morphology. This is particularly clear as regards climatic factors in the case of cultivated plants, where competition has been eliminated. It is true that even in this case there may be found some individuals or strains that are characterized by greater tolerance to external conditions than others, but they undoubtedly represent separate races or incipient species possessing their own specific tolerance. In confirmation of this we may cite the opinion of White (1926, 1928), who considers that a species whose area embraces different climatic zones without any doubt consists of several races differing in their tolerance to temperature conditions. This, in his opinion, accounts for the fact, known to horticulturists, that plants brought from the northern part of the area of a species are better able to withstand cold than those brought from the southern part of the area. The differences between these races are not manifested in the morphological structure of the plants, or, if they are so manifested, the differences are so slight that they are not detectable by the naked eye. White concludes that the ability to withstand low temperatures, in the case of species not accustomed to such temperatures, may be ascribed exclusively to the genesis, by mutation, of cold-resistant races.

As a second indirect proof of his theory Good (*l.c.*, p. 168) cites the relation of plants to edaphic conditions:—

"Many plants are markedly calcifuge and it is usually impossible to find individuals of the same species which do not conform to this character. Nor will cultivation alter this relation unless it results in the production of new forms with new tolerances, so that the original tolerance and species is hidden or lost. Here there is indeed a distinction between cultivated and wild plants, since the effect of cultivation may cause, in a comparatively short time, an evolutionary change which, in nature, would have been much longer delayed".

Good's theory of the migration of floras, though providing a plausible explanation of the movements of floras from one geographical zone to another, leaves out of account two exceptionally important factors: first, the process of divergence which a species undergoes during the course of its migrations and, second, the movements of floras on mountain slopes from one altitudinal belt to another. When a flora moves into a more southern zone, its original place is occupied not only by a flora moving southward from a more northern zone but also by a flora descending from the mountains situated within the limits of the initial zone. These latter elements, as they increase in number, occupy a dominant position. This exceptionally important factor in floral succession is entirely disregarded by Good and, in general, is accorded very little attention. Vegetation belts on mountain slopes have undoubtedly existed during all geological periods and within the limits of all climatic zones of the earth, just as they now exist on mountains in the frigid, temperate, and torrid zones. In our opinion the replacement of floras of lowlands by species of mountain slopes, whenever cold climatic conditions have extended their sway over wider territories, has played a very important rôle in the formation of present-day floras, a circumstance which makes necessary considerable alterations in the migration theory of floral succession.

Another important point. It has usually been assumed that migrations of floras took place only from north to south or from south to north. However, it is now considered definitely established that during the entire Tertiary period and also later there occurred migrations of species from east to west, from eastern Asia into Europe, at first along the mountain systems and later, as the climate became still cooler, also over the plains, replacing the Tertiary flora of Europe that had become extinct.

We may supplement the foregoing by a number of arguments of a different kind. We have seen that, in case of species migration induced by changes in habitat conditions, the plants occupying that part of the area most subjected to the action of these changes find themselves in disharmony with the new conditions of life. They are forced to migrate, following the direction of the climatic change. Such migration will be possible only for the new generation, the seeds of the parent plants, which latter are condemned to remain stationary. What becomes of these plants? Though we know of numerous instances of the dying out of such plants, we also know of a considerable number of cases of their preservation in the most favorable spots of their original area. In the latter cases relic habitats are created, and there arise isolated fragments of an area, to which phenomenon we have already devoted considerable attention.

These habitats may sometimes be of exceptionally limited extent, comprising, for instance, a single mountain or canyon. A certain kind of mountain rock, *e.g.*, chalk cliffs, may constitute the only refuge of such relic plants. Within the limits of their restricted habitats these species will continue to be propagated by seeds, if they have retained the ability to produce them, or vegetatively, if they have lost this ability. These "left-over" plants mix with the newcomers and form one of the elements of the new flora of the given district.

The question arises: Is it possible, by studying the biology of these remnants of a former flora, to elucidate their past history and determine the changes that they have undergone in their habitat? To this question we may reply affirmatively, since various factors in the life of these plants provide definite clues to the knowledge desired.

Periodicity in the Growth and Development of Plants: — That vegetative processes, and also flowering and fruiting, succeed one another in regular fashion is a well-known phenomenon, regarding the causes of which there exists a rather extensive literature (*see* DRUDE, 1913). This alternation of vegetative and reproductive processes, this periodicity in the passing through of the various phases of development, is inherent to all plants, at least all higher plants, regardless of their place of abode on the globe. Even in the tropics, despite the uniformity of climatic conditions, the vegetation is characterized by this rhythm of development. Thus, SCHIMPER (1898) pointed out that in a tropical forest, notwithstanding the fact that climatic periodicity is practically absent, there exists a periodicity in the life of the trees, expressed in the falling of leaves at definite intervals and in an alternation, even in the case of evergreen species, of periods of intensive growth and periods of dormancy.

We shall not go into the question (still a matter of dispute) as to whether periodicity in the life of a plant is primary in character, *i.e.*, independent of external conditions, which may only direct it to one or another side (SCHIMPER, 1898; VOLKENS, 1912), or whether it is the result of external conditions, since absolute invariability of climatic conditions during the entire year does not exist anywhere on the globe (KLEBS, 1911). For our purposes a solution of this problem is not vital. For us it is of importance, first of all, to establish the fact of periodicity of phenomena in the plant kingdom and the existence of interrelations between this periodicity and climatic conditions; and, secondly, to establish the fact that the inherent periodicity of a plant is in most cases hereditarily fixed and not readily subject to change with a change in climatic conditions.

In this connection of particular interest is an experiment made by DIELS (1918) with a number of different species of forest herbs. These herbs were grown in a greenhouse under conditions of uniform temperature (in contrast to the variable climate of their ordinary habitats). The herbs could be classed in three groups, according to how their dormant period was affected: In the first group (type—*Asperula*) the dormant period was completely eliminated; in the second (type—*Leucojum*)—partially eliminated; and in the third (type—*Polygonatum*) —fully preserved. These reactions clearly reflected the geographical relationships of the plants participating in the experiment. For instance, *Asperula*, upon being transferred to greenhouse conditions, no longer had a dormant period and grew without interruption, thus showing its relationship to tropical genera of the same family. The genus *Leucojum* under greenhouse conditions did not enter a dormant period in the fall months but retained its summer dormant period, thus indicating its Mediterranean origin. Lastly, *Polygonatum* retained its normal dormant period corresponding to the climatic conditions of Central Europe, testifying to its holarctic origin.

Consequently, any lack of correspondence between climatic periodicity and the periodicity of the phasic development of a plant is necessarily reflected in some sort of disharmony between the latter and its habitat conditions. This problem has been subjected to detailed investigation by SCHARFETTER (1922). He found that the diversity of periodicity could be classified into two main groups. To the first group belong species the periodicity of whose life cycles corresponds to the periodicity of the Central European climate, so that a plant utilizes all the time suitable for its development. For instance, the apple tree has its dormant period in the winter, flowers in the spring, then develops leaves, and toward autumn begins to fruit. Climatic and life-cycle periodicity correspond completely. To the second group belong species the periodicity of whose life cycles is in disharmony with the climatic periodicity. As an example we may take *Colchicum autumnale*, which flowers late in the fall, when the ovaries formed are most likely to be destroyed by the winter frosts, and the fruits of which ripen only in the spring of the following year, after which this plant, in contrast to most others, enters into a dormant period for the entire summer. Here, in all probability, is reflected the periodicity of the climate of the Mediterranean Region, characterized by a vegetative period with two breaks, the first in the summer, due to lack of rainfall, and the second in the winter, due to low temperature.

Hence, the conclusion may be drawn that, if climatic and life-cycle periodicity correspond, the home of any given plant is the country in which it is found and in which its habitat conditions have not undergone change. If, on the other hand, such correspondence is lacking, the natural conclusion will be that the given plant arose under other climatic conditions. SCHARFETTER considers that in this case the home of the plant should be sought in another place and the plant itself be regarded as an immigrant in the given region. This conclusion, however, is not obligatory, since the indicated disharmony in periodicity might be due to changes in climatic conditions *in situ* or to changes in the plant itself.

This retention by some plants of a periodicity out of harmony with the general rhythm of the surrounding vegetation creates a situation such that these plants do not find themselves in competition with the other species in the same plant community. This gives us a clue to the cause of the preservation of these species in places formerly constituting part of their potential area but now lying outside it. The fact that there exists such a diversity of periodicity among species belonging to the present-day flora of Europe may easily be explained by the great climatic changes that have time and again occurred on that continent, inducing numerous migrations of its species.

Disharmony between the Biological Peculiarities of Some Species and Their Present-Day Habitat Conditions: — In many floras in different regions of the globe, particularly in Europe, there may be found a considerable number of species, the biological peculiarities of which are not in harmony with their habitat conditions. This disharmony constitutes yet another proof of the fact that climatic changes have taken place that have caused migrations of floras, occasional representatives of which have remained on the territory of the original area without having become adapted to their new habitat conditions. In most cases such plants will eventually be crowded out by those species which, as regards their biological peculiarities, find themselves in complete harmony with their habitat conditions.

As a characteristic example of such a plant we shall take the genus *Cyclamen,* which comprises 20 species, distributed primarily in the Mediterranean Basin, particularly its eastern part. This genus is undoubtedly of ancient, Tertiary origin, this being confirmed by its marked isolation from other genera. Everywhere it grows it has preserved with remarkable constancy all its biological peculiarities, which fully harmonize with Mediterranean conditions but which, for many of its species, are not at all in accord with their present habitat conditions. Most *Cyclamen* species flower late in the autumn, when winter is nearly at hand, and in spring or early summer, when the surrounding vegetation is at the apogee of growth, these species lose their foliage and remain dormant for several months, until autumn is well under way, when they enter again into a period of development. Such a life cycle is fully comprehensible under conditions in the Mediterranean Basin with its hot, dry summers and warm, rainy autumns, enabling the vegetation that has become dry and dusty during the summer months to revive and cover itself with fresh leaves and flowers. But under conditions in central Europe, with its rainy summers and cold winters, such periodicity in the life cycle of a plant is not at all in accord with the climatic periodicity. This summer dormant period is not so marked in *Cyclamen europaeum,* which does not shed its foliage and flowers in the summer. In general, however, this periodicity has been preserved even under conditions of cultivation; it has been overcome by age-long cultivation only in the case of *C. persicum.*

As another example of disharmony between the periodicity in the life cycle of a plant and climatic conditions, we may take the ivy (*Hedera Helix*), whose distribution parallels approximately that of the beech. It flowers in September, in northern regions occasionally beginning in August, and in some places continues in bloom as late as January, *i.e.,* during the very coldest months of the year, its fruits ripening, for the most part, only towards spring. Consequently, in nature, flowering of the ivy is rarely observed and then only in warm, protected places; under cultivation, however, it occurs with considerably greater frequency. The ivy is, undoubtedly, an ancient Tertiary plant, which became established in Europe in the Miocene stage, when the winters did not cause such a marked break in its development. Although it has become adapted to cold weather, having penetrated in post-glacial times even into Scandinavia, it has preserved the same periodicity as regards its flowering, a periodicity in complete disharmony with the climatic conditions prevailing in the greater part of its present area (CHEVALIER, 1927).

Another plant which, as regards its time of flowering, likewise does not find itself in accord with present-day habitat conditions is *Colchicum Biebersteinii*, distributed in Asia Minor, on the Balkan peninsula, in the southern part of the U.S.S.R., and in the Crimea. The genus *Colchicum* has been divided into two subgenera, the more ancient *Archicolchicum* and the more recent *Eucolchicum* (STEFANOV, 1926). *Colchicum Biebersteinii* belongs to the former, *i.e.*, the more ancient, subgenus. In the Crimea occasional specimens are found both in steppe and mountain districts, on steppe and sandy soils. The cause of its scarcity here lies in its very early flowering, beginning in January, *i.e.*, in the very coldest month of the year in the Crimea, and ending in March, when the flowering of other Crimean plants has only just barely begun.

The family *Ericaceae*, which has undoubtedly been preserved in Europe since the Tertiary period, is characterized by the xerophytic structure of its leaves, corresponding to the character of its habitats in regions with a winter period of rainfall and in mountainous districts but not at all in accord with its habitats in swampy localities. For instance, the common north European heather, *Calluna vulgaris*, has leaves with a structure markedly xerophytic in character. This species is now very widely distributed, ranging from Spain, the Azores, and Morocco to Norway, Lapland, and West Siberia. In the U.S.S.R. this species extends as far south as Tula Region, being distributed, however, chiefly in the northern regions. Fossil deposits of this heather have only been found beginning with the Ice Age, so that we know nothing regarding its distribution and migrations prior to the Quaternary period. Nevertheless, it is very clear that its xerophytism is a fixed peculiarity of structure that originated under entirely different habitat conditions than those now prevailing over the greater part of its area.

Another type of disharmony we find in *Erica carnea*, belonging to the same family and growing in the alpine zone of western Europe. Its inflorescences develop only toward the end of summer, so that its flowers are not ready for opening before the onset of cold weather, which hinders the further development of the plant. It remains dormant, under the snow cover, until spring, when, with the coming of warm weather, it renews its development and its flowers open. This species, closely related to *Erica mediterranea*, in the Tertiary period undoubtedly did not have such a break in its development, since climatic conditions were more uniform at that time (CHEVALIER, 1927).

The genus *Vaccinium*, most of whose representatives grow in the tropics, has a few species that grow in temperate and subarctic zones of the northern hemisphere, both in the Old and New Worlds, these species no doubt being remnants of a Tertiary, probably mountain, flora. The leaves of most species of *Vaccinium*, even of our north European species—*V. nigrum*, *V. Vitis idaea*, *V. uliginosum*—or of the mountain species, *Arctostaphylos Uva-ursi*, which extends to an altitude of 2,000 m., and also of species of *Oxycoccus*, are thick, coriaceous, and evergreen, in no way corresponding to the foliage of most other plants of northern zones. That these are remnants of a Tertiary flora is testified to by the finding, in Pliocene deposits in central France, of leaves identical in appearance to those of our contemporary blueberries (*Vaccinium uliginosum*), which to-day still grow on these same mountains in France at an altitude of 1,500–1,600 m. (LAURENT, 1904–1905).

Daphne Julia, a Tertiary relic which survived the Ice Age and still grows in the steppes of Voronezh Region, after a rainy summer retains its leaves until the next vegetative period. Hence, the evergreen nature of the genus *Daphne* has not been entirely lost by *D. Julia*, though it is entirely alien to the steppe vegetation and to the climatic conditions now existing in these steppes (KOSO–POLJANSKY, 1928).

An analogous case of physiological atavism we find in the European species of oak and beech (*Quercus Robur*, *Q. sessiliflora*, and *Fagus sylvatica*). It has long since been established that some specimens of these trees, particularly of these species of oak, retain their foliage (in a dead state) until the spring of the following year. This gave grounds for distinguishing between "summer" and "winter" forms. How can we explain this retention of leaves, when all other deciduous species lose their foliage as soon as the weather turns cool in autumn? The only explanation we can give of this phenomenon is that these species in past times dwelt in a different, warmer climate, not characterized by such marked periodicity as now prevails in the northern latitudes, and they were, therefore, enabled to live as evergreens, *i.e.*, not losing their foliage all at once at a definite period but only gradually, a few leaves at a time, simultaneously developing young leaves to replace the old. The correctness of this assumption is confirmed by anatomical data. Usually in deciduous species there may be observed toward autumn, below the points where the leaf-stalks are attached, the development of cork (suberose) tissue and of a special, loose layer of cells, as a result of which the leaf and petiole, separated from the branch (to which it formerly was firmly attached) solely by this cork tissue, easily drop off. The place of leaf abscission is already in advance covered by this layer of suberose cells. Thus, while ordinarily defoliation is a purely mechanical process, due to the suberization and dying away of tissues at the point of leaf abscission, in the above-mentioned species of beech and oak the place of abscission remains alive and green during the entire winter.

Further evidence of the above assumption is the fact that the time of the falling of the leaves in these species of beech and oak coincides with the development of new leaf-buds, as is the case in evergreen plants. In the latter the leaf falls at the precise moment that the bud in the leaf's axil begins development. These species of oak are closely related to evergreen species, and we have full grounds for the assumption that the above-mentioned biological peculiarities show that these species in the past were adapted to different climatic conditions. This is further confirmed by the fact that in southern Europe *Quercus robur* and *Q. sessiliflora* sometimes retain their leaves in a green state throughout the winter. Hence, even after a period of thousands of years, these species of oak and beech, as regards biological periodicity, have not yet become adapted to the climatic rhythm in their present habitats (MAGNUS, 1913).

In this connection it is of interest that the American plant anatomist, Dr. Jeffrey (1917, Chap. XVII), in discussing the hypothesis of recapitulation, gives among other examples the following: —

"In a conifer like the larch, which is differentiated in habit from the mass of the group by its deciduous foliage, we find in the seedling that the leaves persist for several years, thus revealing the probable ancestral condition for the genus. An additional example among the dicotyledons is supplied by the oak. The adult in northern oaks is characterized by deciduous leaves. Oak seedlings and saplings, however, even in the case of typically northern species, retain their leaves during the winter, thus recalling a situation characteristic of the live oaks of warmer latitudes which have evergreen foliage and represent anatomically the primitive type of organization" (p. 235).

Species Not Having a Full Cycle of Development: — In some plant communities it is not uncommon to find species that do not pass through a full cycle of development, *i.e.*, they vegetatively develop normally and flower but do not bear fruit, or they lead a purely vegetative type of life bearing neither flowers nor fruits or, lastly, they do not attain even a normal vegetative development. Such species show clearly that their origin is alien to that of the plant community of which they now form a part. For instance, on meadows in the U.S.S.R. such sedges as *Carex vesicaria*, *C. gracilis*, etc. vegetatively develop normally but do not flower. The same holds true for the steppe cherry of the U.S.S.R., when growing not on the steppes but a little farther north in the southernmost part of the zone of dense oak woods. Similarly, *Caltha palustris*, a plant growing ordinarily in swamps or near bodies of water, is found occasionally in meadows, where it does not bear flowers and even vegetatively develops poorly, the fact that it grows here being evidence that these meadows were formerly swamps. Deciduous, light-loving plants, herbaceous species that develop normally only under conditions of ample sunlight, are sometimes found in the undergrowth in dense spruce forests, where they cease altogether to flower and even have a poor vegetative development.

The fact that in the flora of Spitzbergen there are a considerable number of plants that fail to produce seeds gave grounds for Andersson (1910) to consider that this flora developed under more favorable climatic conditions, warmer than those at present prevailing, and, hence, that it does not now have optimum conditions for growth and development. *Vaccinium uliginosum* grows in Novaya Zemlya, but it neither flowers nor bears fruit there (Regel, 1935). *Rubus arcticus* is found in many plant communities in Finland, but either to the north or south of the optimum zone (62°–65° N.), despite normal flower development, it often fails to bear fruit, being propagated vegetatively (Saastamoinen, 1930). The same phenomenon has been observed in the U.S.S.R., where the optimum zone lies between 60° and 64° N. (Rozanova, 1934). Černjavski (1937) has reported that *Prunus laurocerasus* is found in beech woods north of Vlasina (Jugoslavia), but that it does not bear fruit there.

Extinction of Species: — If changes in climatic and other habitat conditions actually induce migrations of species, then any plants of such species remaining in the original area are subject to extinction. There are numerous facts confirming this. Paleobotany is replete with instances of the extinction of species as a result of climatic changes. A good example is that of the finding of fossil remains of subtropical plants in places located beyond the arctic circle and covered with eternal snow. Moreover, even in our present-day flora we may observe such a process of species extinction induced by changes in climatic conditions.

Braun-Blanquet (1923) has given us a most interesting account of the process of extinction (still in progress) of arctic plants in the mountains of southwestern Europe. The occupation of these habitats by arctic species took place during the Ice Age, when a lowering of the temperature in Europe and the advance of the glaciers southward brought about a corresponding movement of the arctic vegetation. The subsequent increase in temperature, as a result of the regression of the glaciers, was accompanied by an analogous movement of the vegetation. But a few of these arctic species remained in the West-European Alps outside their potential areas. Gradually, as the climate became more arid, there occurred an ever greater decrease in the number of stations of these species. Thus, Braun-Blanquet (*l.c.*, p. 164) cites authorities for the fact that *Ligularia sibirica* had by 1878 disappeared from its former habitat in the Massif Central of France and that *Saxifraga hirculus*, formerly distributed near Nantua, was, in 1897, no longer found there. The species *Oxycoccus quadripetalus*, *Caltha palustris*, *Liparis Loeselii*, *Rhynchospora fusca*, *R. alba*, and *Eriophorum angustifolium* became extinct as a result of the drying up of peat swamps near Lake Bientina in Tuscany. Thanks to the disappearance of the first-named species, the genus *Oxycoccus* altogether disappeared from the flora of central Italy, being preserved only in northern Italy on the southern slopes of the Alps. In a similar way the limits of distribution of *Rhynchospora alba* and of the genus *Liparis* have shifted to the north, for they are no longer found south of the Po.

Analogous instances of the shifting to the north of the limits of distribution of species that had spread their range to the south during the Ice Age are found in Germany. There are data to show, moreover, that the process of extinction (in southern regions) of these species was not recently initiated but began in prehistoric times. This has been confirmed by the finding of fossil remains of such northern species in central and southern Europe. For example, Andersson reported that in swamps adjoining Lago di Garda there were found a large number of fossilized fruits of *Najas flexilis*, which no longer grows in Italy.

In recent years the pollen-statistics method, whereby the pollen contained in samples of peat is analyzed to determine the species from which it was derived, has shown that such samples contain pollen of numerous species that no longer grow in the vicinity of the peat swamps investigated, and it has thus provided much valuable testimony as to the migration and extinction of species.

Climatic Changes in the Tertiary Period and the Resultant Shifting of Vegetation Zones: — Shifting of the climatic zones of our globe has occurred time and again. A study of the fossil remains of plants and their geographical distribution and of the areas of the species making up present-day floras gives ample evidence of this fact. In order to understand the history of our contemporary floras, of greatest significance are those climatic changes which took place during the Tertiary period and during the glaciation of the northern hemisphere at the beginning of the Quaternary period.

The transition from the Cretaceous to the Tertiary period was accompanied in many parts of the globe by the uplifting of mountain chains and the elevation of submerged land above the surface of the sea. Precisely at this time there were formed the greatest of our mountain systems: the Alps, the Himalayas, the Caucasus Mts., and the North American Cordilleras. During the entire Tertiary period there occurred changes in climatic zonation, leading to corresponding changes in the vegetation zones. The most striking evidence of the grandiose changes that took place during the Tertiary period are the remarkable findings of Cretaceous and Tertiary fossil remains in the Arctic. Where now stretch the arctic tundras there formerly grew deciduous forests associated with a temperate climate.

The finding in the Arctic of fossil remains of many genera now distributed in the temperate zones does not, however, suffice to prove that in the past tropical climatic conditions prevailed in the Arctic, as HEER has presumed. That they do not so suffice is confirmed by GOTHAN's report (1915) that he found in the wood of Cretaceous conifers distinct annual rings (found also subsequently in fossil Tertiary wood), which testifies to the existence in the Cretaceous period of alternating cold and warm seasons. In this connection BERRY's conclusions are of interest (1930, p. 29):—

"There is no unequivocal botanical evidence of tropical or subtropical climates at any time in the Arctic. There is no evidence from paleobotany of a lack of climatic zonation at any geological period from which fossil plants are known, although at such times the evidence points to a relative mildness and a lack of sharp zonation, as compared with the present".

This viewpoint of BERRY's is supported by KRYSHTOFOVICH (1929, 1933), on the basis of his numerous investigations of the fossil vegetation of Asia, particularly his study of a collection of fossil plants from the Lozva River in the northern Urals (61° 10′ N.). The species composition of the latter flora has led him to believe that this flora must have required for its existence a mean annual temperature of about 10° C. He further assumes—based on the finding in Eocene and Oligocene deposits in the southern part of European U.S.S.R. (and even more in western Europe) of fossil remains of such plants as *Nipa* and *Sabal* palms—that at 50° N. in southern Russia the temperature must have been not less than 18°–20° C., while in western Europe on the same parallel it might have been still higher, corresponding, therefore, to subtropical or even tropical climatic conditions.

As regards the Tertiary flora of Asia, KRYSHTOFOVICH states the following (1929, p. 307):—

"Turning now to the Tertiary Siberian floras we are not able to find in them any remains that would permit us to suggest the existence anywhere in this country of strictly tropical or subtropical conditions during the Tertiary period. On the contrary all the known facts concerning the territory which stretches from Turgai and Tomsk on the west to Vladivostok, Corea and Sakhalin on the east, demonstrate the former distribution there of a flora composed mostly of temperate types such as *Fagus, Ulmus, Alnus, Betula, Cory-*

lus, *Populus*, *Juglans*, *Comptonia* and *Trapa*, and almost devoid of evergreen elements showing a southern character; or the southern plants, if present, are rare and doubtful. There are no palms, nor are there cinnamons or figs of a tropical type such as are so conspicuous in the older Tertiary flora of Europe. . . . . In proceeding farther east we find the same monotonous Tertiary flora passing across the Pacific into Alaska . . . However, whilst in Siberia proper any traces of a former much warmer climate and an associated flora are lacking, some traces of these phenomena are found in Turkestan and in Japan".

Nevertheless, there is no doubt that climatic conditions in Siberia, even as far north as the Arctic Ocean, were during the Tertiary period considerably milder than at present, which made possible the existence of deciduous species resembling types at present found in eastern Asia and the eastern part of the United States. At the same time, beginning with the Paleocene stage, in western Europe, the Ukraine, and the central part of Russia in Europe there was distributed an evergreen tropical and subtropical flora, characterized by such evergreen plants as the *Nipa* and *Sabal* palms and species of *Cinnamomum*, *Ocotea* (*Oreodaphne*), and other genera of a tropical type. This flora, according to KRYSHTOFOVICH (1932), resembles most closely the Indo-Malayan type of paleotropic flora, whereas the later Tertiary flora of these regions assumes an aspect similar to the present-day flora of eastern Asia and the eastern part of the United States.

Still later, at the very close of the Tertiary or the beginning of the Quaternary period, due to the further lowering of temperature, the covering of a large part of Europe and northern Siberia with glaciers, and the considerable decrease in precipitation in the Mediterranean Basin and also in eastern Europe and in northern and central Asia, the species composition of the floras of the regions we have been discussing completely changed. The vegetation of the tropical zone, on the other hand, developed almost undisturbed, due to the fact that habitat conditions there remained almost unchanged. Similarly, the vegetation of southeastern Asia, particularly Japan and southwestern China, and of northern Mexico and the southeastern and southwestern sections of the United States has remained practically unchanged since the Tertiary period, not having undergone those sharp climatic changes that the vegetation of Europe and the northern part of North America has repeatedly had to undergo.

To explain these great climatic changes many theories have been advanced. It is not our purpose here to go deeply into this problem; for us it suffices merely to point out that the vegetation of the earth has, at various times during the earth's history, been subjected to considerable changes, corresponding to changes in climatic zonation, and that these changes have been reflected not only in the species composition of floras but also in the numerical distribution of species.

The remarkable evolution of the plant world has not been a uniform and gradual process. On the contrary, everything indicates that the sudden changes in habitat conditions, caused by violent upheavals of the earth's crust, that have occurred time and again during the earth's long history have given a marked impetus to the evolution of forms, have induced sudden mutational changes leading to the abrupt creation of new forms, to processes of accelerated species-formation, and also to migrations of floras. If the earth had remained all the time in the same state as, let us say, in the Carboniferous period, with the same distribution of climatic zones, the same climatic conditions, and the same distribution of land and sea, there is no doubt that the earth's vegetation would not have attained, simply by the slow processes of natural selection, that stage in its evolution at which it now finds itself. Instead of the angiosperms, having an anatomical structure adapted to intense light and moderately humid habitat conditions, perhaps even to-day there would still be flourishing *Lepidodendron* and *Sigillaria*, and the animal world would be represented not by mammals and man but by gigantic reptiles.

The great geological revolutions that have periodically occurred, inducing intensified mountain-formative processes and accompanied by the advance and regression of seas and the shifting of continents as regards their position in relation to the poles, have served as the cause of changes in the location of the climatic zones. These latter changes, in turn, have led to ever new changes in the distribution of plants and animals, producing an enormous effect on their development and evolution.

During long ages, extending to the very close of the Tertiary period, the distribution of the climatic zones favored the development of the vegetation of the northern hemisphere. In contrast to the comparatively uniform development of the flora of the northern hemisphere, the development of the flora of the southern hemisphere was time and again disturbed. Chief among such disturbances was the great southern glaciation in the Permo-Carboniferous epoch (perhaps also a second later one), the principal center of development of which lay in South Africa. These glaciations, as well as subsequent climatic changes during the Quaternary period, constitute one of the reasons for the poverty of the flora of tropical Africa as compared with Asia and South America.

A second cause of the more uneven course of development of the vegetation of the southern hemisphere was the separation of the continents (accompanied by a decrease in rainfall), which began as far back as the Mesozoic era. This was necessarily reflected in the number of species in the flora of Africa, since, besides being cut off from the vegetation of America, Africa was almost cut off from the rich vegetation of Europe and Asia, due to the existence in the region of the Sahara in the Cretaceous period of a great sea and in later times of a great desert that barred the way to species migrating from the north.

In the early part of the Quaternary period the fate of South Africa was shared by Europe and North America, which continents had previously been characterized by an exceptional wealth of plant forms. At the present time Europe is one of the poorest continents as regards the number of species in its flora. This may be ascribed to the fact that the flora destroyed during the Ice Age had here, as compared with North America, very little possibility of becoming restored. The stocking of this continent with new vegetation could proceed only from the south, from the Mediterranean Basin, and from the east, from northern and southwestern Asia, which regions at that time also had greatly impoverished floras. The path from the west, from North America, was closed, since the Atlantic Ocean had already been formed;

that from Africa was also closed, since the land-bridges that had formerly connected Europe and Africa were no longer in existence.

The Ice Age and Its Effect on Vegetation: — The climatic changes that commenced about the middle of the Tertiary period led to the Quaternary glaciation of the northern hemisphere. Having first started, presumably, in North America, it spread over Europe and, to some extent, over northern Asia. Considerable expanses of territory in the latter continent, due to the more continental character of the climate, remained uncovered by the glaciers. In all probability the present climatic conditions of Greenland closely approximate those prevailing over a considerable part of the northern hemisphere many thousands of years ago. Glaciers covered all of western Europe, extending as far south as the Thames, on the west, and the Carpathians, on the east. All of northeastern Europe was completely covered by the great ice sheet, which extended along the valleys of the Dnieper and the Don as far south as 49°–50° N. In Siberia, on the other hand, it did not cross the 60th parallel, and a number of regions, particularly in the northeastern part, remained uncovered by the ice-sheet. The extension of the ice-sheet was not uninterrupted: now it advanced, now it retreated. Glacial epochs were followed by interglacial epochs, characterized by comparatively mild climatic conditions. The vegetation destroyed by the glaciers was gradually restored as a result of the return of some of the species that had been forced to migrate to the south. As has been shown by various collections of fossil plants, including those made at Hötting (near Innsbruck) in the Tyrolian Alps (WETTSTEIN, 1892), there grew in these regions during the last interglacial epoch many plants not found there at the present time. One of the chief stations of the chestnut was located at an altitude of 1,080 m. above sea level; grapes grew in the same places as now; *Buxus* and *Rhododendron ponticum* were widely distributed, whereas now in the Mediterranean Basin there have survived only a few relic specimens in isolated localities; *Laurus canariensis*, *Ficus carica*, and *Cercis Siliquastrum* grew near Paris; fossil remains of *Euryale ferox*, an aquatic plant very closely related to *Victoria regia*, now distributed in India and eastern Asia, were found in central Russia near the Oka River (SUKATSCHEFF, 1908).

The glaciation of enormous expanses of territory in the northern hemisphere could not but be reflected in a lowering of temperature in other parts of the globe, even including the tropics, which led to the formation there of a snow cover—or, if such already existed on mountain tops, to a lowering of the snow-line—and to an increase in precipitation throughout the southern hemisphere, *i.e.*, to so-called *pluvial periods*.

A study of glacial phenomena has shown that the glacial epochs did not constitute catastrophes that suddenly descended upon the plant and animal world of the continents. The advance and retreat of the ice occupied a period of thousands of years, during which, as a result of changes in climatic conditions, plants and animals migrated, as the ice advanced, from the north to the south and from the upper altitudinal zones to the plains and valleys, and, as the ice retreated, returned, in part, to the north and to the higher altitudes.

Refuges of Species during Glacial Periods and Migrations of Species: — The striking fact that the same species are found in the Arctic and on mountain peaks, in localities separated from one another by the lowlands of the temperate zone, where these species cannot grow, was explained by FORBES and DARWIN as an effect of climatic conditions during the Ice Age. As the great ice-sheet advanced southward, the arctic flora also migrated in the same direction, taking the place of representatives of the flora of the temperate zone that had died out or had likewise migrated. During the interglacial epochs, as the climate became warmer, the arctic species retreated to the north and to the mountain tops.

During the glacial epochs the species of the north temperate zone migrated southward, following chiefly the great mountain systems, into southern Europe, southern Asia, Africa, South America, and that part of North America south of the ice-sheet, *i.e.*, to the unglaciated regions, where pluvial periods coincided in time with the northern glaciations. During the dry, warm, interglacial epochs, on the other hand, desert xerophytes spread from Africa northward into the temperate zone.

After the final retreat of the ice-sheets and the shrinking back of the snow-line the mountain slopes and enormous expanses of territory, on which the former vegetation had been destroyed, began gradually to be re-stocked with plants, which spread northward following close in the wake of the receding ice. The mixed flora, composed of arctic species that had previously migrated southward and were growing on the fringe of the ice-sheet and of alpine species that had descended from the mountain tops, now returned to the north and to the alpine heights, inhabiting the territories recently freed from ice.

In mountainous regions located at more southern latitudes, where the vegetation was affected to only a slight degree by glacial phenomena, and also within the limits of the Alps themselves, where undoubtedly even in regions covered by glaciers there were cliffs, canyons, and sheltered spots that remained free of the ice covering, the pre-glacial flora was preserved (*see* ENGLER, 1916, and BRIQUET, 1908) and there took place an intermingling of the newcomers and the indigenes. An analysis of the present species composition of the flora of the Alps reveals with all clarity its mixed character (JEROSCH, 1903).

As the climate became still warmer and the fringe of the retreating ice-sheet withdrew farther north, the mixed arctic-alpine flora was crowded farther and farther north by the forest flora, the species composition of which had changed, as we have seen, corresponding to the climatic changes that had taken place. The freed territory was likewise colonized by herbaceous plants. Paleobotanic data give us practically no aid in judging as to the composition and origin of the elements of this flora. We needs must resurrect the history of its formation on the basis of an analysis of the species composition of the flora occupying this territory to-day.

Whence came this vegetation on territory covered with ice for thousands of years? There is no doubt that the re-stocking of this territory did not occur all at once but at various times and that the invading species came not from one but from many places, from many centers of dispersal. As such centers there must have served, first of all, those territories and mountain peaks (so-called "nunataks") which, although lying within the glaciated areas, were not covered by the ice. That pre-glacial vegetation did survive in such places may now be considered a definitely established fact.

Secondly, as centers of dispersal there served those territories lying directly south of the ice-sheet's fringe. We now regard as erroneous the opinion formerly held that during the Ice Age the vegetation was destroyed not only on the glaciated territory but also in a wide radius about it. There are data establishing the fact that the arctic (*Dryas*) flora occupied only a more or less narrow strip along the edge of the ice-sheet, beyond which the vegetation comprised herbaceous and forest species of a more heat-loving type. This is evidenced by the fossil tree-trunks and other parts of woody plants found within the limits of this zone. Formerly, often solely on the basis of the fallacious assumption that at such a distance from the ice-sheet there could have existed only tundra vegetation, such fossil remains were referred only to the interglacial epochs.

Thirdly, as centers of dispersal for the re-stocking of the territories freed from the ice, there served those refuges or sheltered stations, primarily mountain systems, where the Ice Age did not have such catastrophic consequences for plant life and where there survived a Tertiary flora, though in some cases in an impoverished state.

There is a great diversity of opinion as to which regions retained during the Ice Age their forest vegetation. The chief, and generally recognized, refuges in Europe may be regarded as the following (from west to east): the Mediterranean Basin, including the Pyrenees, Cevennes, and other mountains of southern France, the Apennines, the southern spurs of the Alps, and the mountain systems of the Balkans; the mountains of southern Germany (possibly also the Schwäbische Jura); the mountains of Lower Austria (possibly including the adjoining Böhmisch-Mährische Höhen); the Carpathians and the Banat hills; the mountains of the Crimea and the Caucasus and the adjoining mountain systems of Asia Minor and Iran (LÄMMERMAYR, 1923). Most of these mountain systems run east and west, in consequence of which they constituted a barrier beyond which the heat-loving Tertiary vegetation could not retreat as it fled from the cold advancing from the north.

The impoverishment of the Tertiary vegetation was a slow process. After each advance of the ice-sheet it lost a number of its elements. Only the southern part of North America and those southern regions of Europe protected on the north by mountain chains, such as the Mediterranean Basin, Transcaucasia, and southern Crimea, retained a large number of species of their pre-glacial flora. Consequently, the vegetation occupying the territory freed from the glaciers was considerably poorer than that inhabiting the same territory in pre-glacial times. In North America, where the mountain chains run north and south, the vegetation that retreated during the Ice Age did not encounter any such barriers as in Europe, and so it was able to migrate considerably farther south. This explains why there is a greater percentage of Tertiary elements in the present-day floras of formerly glaciated territories in North America than in those of similar territories in Europe.

But this return of species to their old habitats does not by any means account for all the migrations of species that resulted from the Ice Age. The territory freed from the ice offered favorable conditions, due to the absence of competition and of a closed vegetation, for the invasion of plants from lands but slightly affected by the Ice Age. Moreover, the alternation, in post-glacial times, of dry and humid climatic conditions led to corresponding migrations of the hydrophytic and xerophytic elements of the various floras. Asia constituted the chief center from which there flooded into Europe a new vegetation. South Siberian and Central Asiatic species, which penetrated into Europe, presumably, south of the Urals, species of the Aral-Caspian desert flora, mountain species of the Caucasus and Western Asia—these were the elements which enriched the flora of Europe and the advance of which to the west is, in all probability, still in progress.

Climatic Changes and Successions of Vegetation; The Post-Glacial Period: — The retreat of the ice-sheets was not achieved at one stroke. They alternately advanced and retreated several times, each succeeding advance being less than the preceding. Similarly, the snow-line receded to higher altitudes, at each stage lying 300–400 m. higher than before. The present line lies at approximately this distance above the ridge of moraine left by the last advance of the glaciers. After the final retreat of the glaciers and ice-sheets the territory previously occupied by them, including all the northern part of eastern Europe, was covered by numerous lakes and the entire Baltic depression was filled with water. These lakes have only been preserved in small part to the present day; most of them, thanks to a gradual deepening of the river beds, have either dried up or become converted into peat-bogs.

Climatic conditions in the post-glacial period, according to Blytt (1876, 1882), were characterized by age-long oscillations, warm, humid periods alternating with dry, colder periods. Assuming such climatic oscillations, it is possible to understand the present-day distribution and grouping of species in the flora, let us say, of the Scandinavian peninsula. In particular, as regards the flora of Norway, Blytt established six groups of species, each having a definite geographical distribution:

The first group is confined in southern Norway to the mountains and in northern Norway to the plains and bears a purely *arctic* character. Typical species of this group are *Dryas octopetala* and a dwarf willow, *Salix reticulata*. This type of vegetation is likewise characteristic of Greenland, Spitzbergen, and other arctic regions.

The second group is composed of species which, like the preceding, avoid humid habitats, growing far from the shore and at altitudes up to 2,000 ft. above sea level. However, this group is considerably richer in species than the preceding group. Here we already find deciduous shrubs and trees, such as the hazelnut, elm, linden, ash, maple, oak (*Quercus pedunculata*), *Sorbus aria*, etc., and also a considerable diversity of herbaceous species. This type of flora Blytt designated as *boreal*.

Still richer in species is the third group, also connected with arid, continental climatic conditions. In it we find a new sub-boreal element, *viz.*: *Spiraea filipendula*, *Geranium sanguineum*, *Thymus chamaedrys*, *Rhamnus cathartica*, *Fragaria collina*, etc.

In addition to these three groups of species, there are two others which, in contrast to the preceding, prefer humid habitats near the seashore. The first of these BLYTT calls an *Atlantic* flora, inhabiting the most humid places along the shore. The number of species belonging to this flora become less and less, the farther from the shore and the more continental the habitat conditions. To this group belong: *Ilex*, *Digitalis purpurea*, *Erica Tetralix*, etc. The second is a *sub-Atlantic* flora, concentrated in the southernmost parts of the shore. Here we find, for instance, *Gentiana pneumonanthe*, *Cladium mariscus*, and *Teucrium scorodonia*.

Lastly, the sixth group comprises the *subarctic* flora distributed throughout the entire country. It also is linked with humid climatic conditions, for its species do not avoid the coast regions.

The species composing these groups are frequently found in mixed communities, but in some places the species of one and in others those of another group predominate to such an extent that there can be no doubt that each group possesses a specific history of origin.

In view of the fact that the Scandinavian peninsula during the Ice Age could have retained only insignificant remnants of its pre-glacial vegetation, since it was almost entirely covered by the great ice-sheet, it is clear that its present vegetation could have penetrated its territory only from places farther south and only after the ice-sheet had begun its retreat. Moreover, each of the above-mentioned floras constitutes a group of species that simultaneously invaded those parts of Scandinavia from which the ice had receded.

If we now keep in mind that these floristic groups constitute combinations of xerophytic or hydrophytic elements, the conclusion naturally follows that the climate of the post-glacial period underwent great changes, periods of cold, dry climate alternating with periods of warm, humid climate. Such alternations of climate occurred repeatedly, and during each respective climatic period there took place an invasion into Norway of a corresponding group of species. Each group of newcomers, adapted to the climatic conditions prevailing at the time, crowded out their predecessors that had invaded the peninsula during the preceding climatic period. The latter, however, did not entirely disappear; some were preserved in sheltered places, where the climatic conditions were such as to secure them from competition with the new invaders.

There remains now only to establish in what sequence Norway was restocked with vegetation, and, consequently, in what sequence there took place the alternate periods of cold, dry and warm, humid climate. The answer to this question we find in the peat-beds of Scandinavia, the plant remains in the successive layers of which testify very positively to a definite sequence in the succession of floras. These peat-beds, according to BLYTT, show that three times during the course of their formation there were dry periods, lasting presumably for thousands of years, when the formation of peat ceased and the peat-bogs became grown over with forests. Alternating with these dry periods there were humid periods, when the forests perished and their place was occupied by bog moss (*Sphagnum*) and the peat deposits again steadily grew over long periods of time.

STEENSTRUP (cited by BLYTT) established in peat deposits in Denmark four layers, corresponding to four stages in the history of the invasion of this region by vegetation. The lowest layer is characterized by the predominance of aspen leaves (*Populus tremula*); above it lies a layer in which debris of pine trunks predominates; above that a layer in which the oak, *Quercus sessiliflora*, predominates; and, lastly, in the uppermost layer we find the alder (*Alnus glutinosa*) to be predominant. He designates these layers, found in the same sequence in many peat-beds in Denmark, as the aspen, pine, oak, and alder layers. The conclusion necessarily drawn is that these four layers in the peat-beds of Denmark correspond, in the sequence of their stratification, to the four layers in the peat-beds of southern Norway. Analyzing these data, BLYTT considers it possible to presume the following sequence in the alternation of climatic conditions in Denmark and southern Norway:—

1. End of Ice Age. Climate humid.
2. Arctic flora: *Dryas*, arctic willows and birches. Climate dry, continental, arctic.
3. Peat with leaves of aspen and birch (*Betula odorata*).
4. Remains of roots and other parts of forest trees.
5. Peat with remains of *Pinus*.
   (During the deposition of the 3rd, 4th, and 5th layers there occurred the invasion of these regions by a *subarctic flora*.)
6. Remains of trunks and other parts of forest trees, including the hazelnut, oak, and other heat-loving species. (Invasion by *boreal flora*.)
7. Peat with remains of the trunks of the oak, *Quercus sessiliflora*, which at that time was more widely distributed than now, attesting a warmer climate. (Invasion by *Atlantic flora*.)
8. Remains of roots and other parts of forest trees. (Invasion by *sub-boreal flora*.)
9. Peat — almost pure *Sphagnum*. (Invasion by *sub-Atlantic flora*.)
10. Contemporary. The peat bogs are, for the most part, dried up and partially overgrown with forests; a new layer of roots is in process of formation, which will again be buried under layers of peat as soon as there ensues a new pluvial period. (*Contemporary flora*.)

This alternation of climatic periods BLYTT associated with changes in the direction of ocean currents, caused by shiftings of the Scandinavian shield and consequent changes in the temperature of the water of the seas washing the coasts of Scandinavia. These views have been confirmed by investigations as to variations in the level of the North and Baltic Seas.

BLYTT's theory was definitely substantiated by the investigations of peat-beds made by SERNANDER (1892, 1910) and his followers, who employed very exact and improved methods of investigation. Basing himself on the fundamental principles of BLYTT's system, SERNANDER gives his own scheme of the alternations of climatic periods, in constructing which he also employs archaeological data. Besides the fact that the alternations in climatic periods are provided with more substantial proof, SERNANDER's scheme has the advantage of being simpler than BLYTT's. This scheme, known as the Blytt-Sernander scheme, with some modifications based on subsequent investigations, is at present the one most widely accepted. However, many other viewpoints on this subject, often quite contradictory, have been expressed. Despite their great interest, we cannot go into them here. The chief cause of the many disagreements was the chance nature of the findings of plant remains and the lack of precision in determining the numerical ratios of the various species in the deposits, this leading to contradictory interpretations. Fortunately, we now possess a more perfected and precise method of determining the species composition of former floras and the numerical ratios of the component species. This is the method of pollen-statistics, which involves a study of the fossilized pollen of plants found in peat deposits.

Method of Pollen-Statistics: — Fossilized pollen grains in post-glacial deposits have been known since the end of the nineteenth century. The first work on a calculation of species ratios based on pollen-statistics was published by WEBER in 1896, and the first methodological principles of pollen analysis were worked out by LAGERHEIM in 1905 and 1909. But a detailed elaboration of the method of pollen-statistics was made only in 1916 by VON POST and his associates, the method being at first applied chiefly in Sweden. It has now become widely adopted, and, as a result, we already have considerable data and maps of the former distribution of forests and the species composing these forests for a number of European countries. At the same time, the method itself has been further elaborated and made more precise.

The method of pollen-statistics is based on the fact that the pollen grains of different species of plants may be distinguished from one another by their specific structure. The structure of the outer coat of the pollen grain is specific for each genus and often for each species, constituting a good character for taxonomic classification. Moreover, the pollen of many species is well preserved in peat deposits, which makes it possible by studying such fossilized pollen not only to determine the genera and species comprised in a forest at the time the pollen was produced, but also, by calculating the number of pollen grains for each species, to determine the relative proportions of the various species in the given forest.

The many investigations of this kind that have been made have already elucidated quite precisely for a number of countries the composition of the vegetation at different stages during the post-glacial period. As an example, we may take RUDOLF's (1931) conclusions, based on a summary of all available data, whereby he divides the history of the forests of central Europe into the following four periods:

1. *Birch-Pine Stage* (*Pre-Boreal Period*). Throughout all of central Europe the chief trees are the pine (*Pinus*), the birch (*Betula*), and the willow (*Salix*). The predominance of the birch, and also of the willow, is greater toward the west and north. Other forest species are sporadically, though in places quite widely, distributed. The spruce (*Picea*) for some time is dominant in central Russia, and it is also widely distributed in the Carpathians and eastern Alps. At times, instead of the birch, the pine becomes predominant.

2. *Hazel Stage* (*Boreal Period*). The hazel (*Corylus*) frequently occupies a dominant position, particularly in the west and in mountainous districts of the east. Oak woods begin to become widely dis-

tributed throughout the entire region. The spruce at first is limited to mountain habitats in the Alps and Carpathians, but it is later found also in the more northern parts of the mountain chains. Only in the western mountainous districts is the fir beginning to become distributed. The birch and pine are becoming less widely distributed.

3. *Mixed Oak Forest Stage (Atlantic Period).* Mixed oak forests occupy a dominant position or, in any case, are characteristic of the forests of central Europe of this period. At first, besides the oak, the elm (*Ulmus*) and the lime or linden (*Tilia*) dominate in the forests; later only the oak. In districts where the spruce and fir are distributed, these trees, which find themselves at a stage of expanding distribution, compete with the oak. Finally, the fir in the central altitudinal belts of the mountains in the west and the spruce in the east and in the higher altitudinal belts of the west predominate, while the oak retains its dominant position in the deciduous forest districts. The hazel has by this time become only a shrub. Toward the end of the Atlantic period the forests of central Europe are enriched by the beech, which becomes widely distributed, together with the hornbeam (*Carpinus*).

4. *Beech Stage (Sub-Boreal-Sub-Atlantic Period).* The beech (*Fagus*) occupies a dominant position in the deciduous forests of central Europe. It is widely distributed in coniferous forests, but here it finds itself in competition with the conifers, particularly the fir.

5. *Stage of Man's Influence on Forests.* The forests are destroyed over a considerable part of the territory of central Europe. Reforestation destroys the virgin state of the forest. In coniferous forest districts deciduous species are crowded out by the spruce and pine, while in deciduous forest districts the distribution of the pine is artificially fostered.

From the hazel stage to the beech stage the upper limit of the forests, as well as the altitudinal boundary of a number of forest species in the mountains of central Europe, was several hundred meters higher than now, corresponding to the more northern distribution of a number of species in Scandinavia during the warm, postglacial period.

This complex picture of the succession of vegetation cannot but be the result of a number of different causes, just as at present the areas of species are the result not of any one factor but of a combination of factors. Moreover, there is no doubt that the chief of these factors is climate, and also that the described succession of plant formations is due primarily to changes in climatic conditions. There no longer exist any disagreements on this point, but the precise determination of the climatic periods, as based on changes in the composition of the vegetation, and their chronological sequence are problems that are far from being solved and that are still the subject of heated dispute.

All the foregoing leaves no doubt as to the significance of historical causes—of the changes in climate and in the distribution of land and sea that have taken place on the globe in different geological periods—for an understanding of the present distribution of species and floras and their groupings. We shall now examine various current theories regarding the historical changes that have occurred on the face of the earth, with the aim of determining which of them, from a phytogeographical standpoint, is the soundest and most plausible.

References:

ANDERSSON, G., 1910: *In* "Die Veränderungen des Klimas seit dem Maximum der letzten Eiszeit" (Leipzig).

BERRY, E. W., 1930: The past climate of the North Polar region (Smithsonian Misc. Collections, Vol. 82, No. 6).

BLYTT, A., 1876: Essay on the Immigration of the Norwegian Flora during Alternating Rainy and Dry Periods (Christiania).

BLYTT, A., 1882: Die Theorie der wechselnden kontinentalen und insulären Klimate (Engler's Bot. Jahrb., Vol. 2).

BRAUN-BLANQUET, J., 1923: L'Origine et le Développement des Flores dans le Massif Central de France (Paris).

BRIQUET, J., 1908: Les réimmigrations postglaciaires des flores en Suisse (Acta Soc. Helv. Sci. nat., Vol. 1).

BROCKMANN-JEROSCH, H., 1914: Zwei Grundfragen der Paläophytogeographie (Engler's Bot. Jahrb., Vol. 50, Supplt.).

BROOKS, C., 1925: The Evolution of Climate (London).

ČERNJAVSKI, P., 1937: Pollenanalytische Untersuchungen der Sedimente des Vlasinamoores in Serbien (Beih. z. Bot. Centralbl., Abt. B., Vol. 56).

CHEVALIER, A., 1927: Biogéographie (*In* MARTONNE "Traité de Géographie Physique", III; Paris).

DARWIN, CHAS., 1859: Origin of Species (London; 6th ed., 1911).

DARWIN, CHAS., 1868: The Variation of Animals and Plants under Domestication, Vol. II (London).

DIELS, L., 1918: Das Verhältnis von Rhytmik und Verbreitung bei den Perennen des europäischen Sommerwaldes (Ber. d. Deutsch. Bot. Ges., Vol. 36).

DRUDE, O., 1913: Die Oekologie der Pflanzen (Stuttgart).

ENGLER, A., 1879: Versuch einer Entwicklungsgeschichte der Pflanzenwelt, Vol. I (Leipzig).

ENGLER, A., 1916: Beiträge zur Entwicklungsgeschichte der Hochgebirgsfloren (Acta Preuss. Ak. Wiss., Ph.-Math. Kl., No. 1).

ERDTMAN, O. G. E., 1927–34: Literature on pollen-statistics (Geol. Fören. Stockholm Förhand., Vol. 49, No. 2; Vol. 52, No. 2; Vol. 54; Vol. 56, No. 3).

ERDTMAN, O. G. E., 1943: An introduction to Pollen Analysis (Chronica Botanica Co.)

GAMS, H. und NORDHAGEN, R., 1923: Postglaziale Klimaänderungen und . . . in Mitteleuropa (München).

GAMS, H., 1927: Die Ergebnisse der pollenanalytischen Forschung in Bezug auf die Geschichte der Vegetation und des Klimas von Europa (Zeitschr. f. Gletscherkunde, Vol. 15).

GAMS, H., 1931: Das ozeanische Element in der Flora der Alpen (Jahresber. Ver. z. Schutz d. Alpenpfl., Vol. 3).

GAMS, H., 1931: Die klimatische Begrenzung von Pflanzenarealen und die Verteilung der hygrischen Kontinentalität in den Alpen (Zeitschr. d. Gesellsch. für Erdkunde, Nos. 1/2, 5/6, 9/10).

GOOD, R. D'O., 1931: Plant distribution (Nature, Vol. 127, No. 3204).

GOOD, R. D'O., 1931: A theory of plant geography (The New Phytologist, Vol. 30, No. 3, pp. 149–171).

GOTHAN, W., 1915: Pflanzengeographisches aus der paläozoischen Flora mit Ausblicken auf die mesozoischen Folgefloren (Engler's Bot. Jahrb., Vol. 52).

HEER, O., 1883: Die fossile Flora Grönlands (Leipzig).

HEGI, G., Flora von Mitteleuropa, Vol. V, Pts. 2 and 3.

HILDEBRAND, F., 1902: Die *Cyclamen*-Arten, als ein Beispiel für das Vorkommen nutzloser Verschiedenheiten im Pflanzenreich (Beih. z. Bot. Centralbl., Vol. 22).

HOERNES, R., 1911: Das Aussterben der Arten und Gattungen (Graz).

JEFFREY, E. C., 1917: The Anatomy of Woody Plants (Chicago).

JEROSCH, M., 1903: Geschichte und Herkunft der Schweizerischen Alpenflora (Leipzig).

KASHKAROV, D. and KOROVIN, E., 1931: An analysis of the ecological routes of dispersal of flora and fauna (In Russian; Jour. Ecol. and Biocoenol., No. 1).

KLEBS, G., 1911: Über die Rhytmik in der Entwicklung der Pflanzen (Sitzungsb. Ak. Wiss. Heidelberg, Vol. 23).

KÖPPEN, W. und WEGENER, A., 1924: Die Klimate der geologischen Vorzeit (Berlin).

KOSO-POLJANSKY, B., 1928: Glaziale Pflanzenrelikte auf dem Orel-Kurskischen Plateau (*In* SCHENCK und KARSTEN, Vegetationsbilder, Reihe 19, No. 1/2).

KRYSHTOFOVICH, A., 1928: Greenland's Tertiary flora in the northern Urals and botanico-geographical provinces in the Tertiary period (In Russian; Nature (Priroda), No. 5).

KRYSHTOFOVICH, A., 1929: Evolution of the Tertiary flora in Asia (New Phytologist, Vol. 28).

KRYSHTOFOVICH, A., 1932: The Tertiary flora of the North Polar region and Wegener's theory (In Russian; Bull. Com. Géol., Leningrad, Vol. 51).

KRYSHTOFOVICH, A., 1933: Fossil flora from the Lozwa River, North Ural (In Russian; Eng. summary; Trans. United Geol. and Prospect. Service of U.S.S.R., Fasc. 291).

KRYSHTOFOVICH, A., 1935: Palms in Tertiary deposits of the southern Urals (In Russian; Nature (Priroda), No. 2).

KUPFFER, K., 1925: Grundzüge der Pflanzengeographie des Ostbaltischen Gebietes (Riga).

LÄMMERMAYR, L., 1923: Die Entwicklung der Buchenassoziation seit dem Tertiär (Fedde's Repert. Spec. Nov. Beih., Vol. 24).

LAURENT, L., 1904–1905: Flore pliocène des cinérites du Pan-de-la-Mougudo (Ann. Mus. Hist. Nat. Marseille, Vol. 9).

MAGNUS, W., 1913: Der physiologische Atavismus unserer Eichen (Biolog. Centralblatt, Vol. 33, No. 6).

MÜLLER, P., 1933: Verbreitungsbiologie der Garigusflora (Beih. z. Bot. Centralbl., Abt. II, Vol. 50).

POST, L. VON, 1916: Skogträgpollen i sydsvenska torvmosselagerföljder (Geol. Fören. Stockholm Förhand., Vol. 38).

REGEL, C., 1935: Die Reliktvereine in der Arktis (Cohn's Beitr. z. Biol. d. Pfl., Vol. 23, No. 2).

RIKLI, M., 1912: Lebensbedingungen und Vegetationsverhältnisse der Mittelmeerländer und atlantischen Inseln (Jena).

ROSA, DANIEL, 1903: Die progressive Reduktion der Variabilität und ihre Beziehungen zum Aussterben und zum Entstehen der Arten (Jena).

ROZANOVA, M. A., 1934: Berry resources of the North (In Russian; *in* "Problems of Northern Agriculture", No. 4; Institute of Plant Industry, Leningrad).

RUDOLF, K., 1931: Grundzüge der nacheiszeitlichen Waldgeschichte Mitteleuropas (Beih. z. Bot. Centralbl., Vol. 47).

SAASTAMOINEN, S., 1930: Die nordische Himbeere, *Rubus arcticus* L. (Ann. Soc. Zool. Bot. Fenn. Vanamo, Vol. 13).

SAKS, V. N., 1936: On the Quaternary glaciation of northern Siberia (In Russian; Arctica, Vol. 4).

SCHARFETTER, R., 1922: Klimarhytmik, Vegetationsrhytmik und Formationsrhytmik (Oest. bot. Zeitschr., No. 7–9).

SCHIMPER, A., 1898: Pflanzengeographie auf physiologischer Grundlage (Jena; 2nd ed., 1908).

SERNANDER, R., 1892: Om de uplandska tormossarnes bygnad (Bot. Notiser).

SERNANDER, R., 1910: Die schwedischen Torfmoore als Zeugen postglazialer Klimaschwankungen (Geol. Fören. Stockholm Förhand.).

SEWARD, A. C., 1931: Plant Life through the Ages (Cambridge; 2nd ed., 1933).

STEFANOV, B., 1926: Monograph on the genus *Colchicum* (Bull. Acad. Sci., Sofia, Vol. 12).

SUKATSCHEFF, W. N., 1908: Über das Vorkommen der Samen von *Euryale ferox* Sol. in einer interglaz. Ablagerung in Russland (Ber. d. Deutsch. Bot. Ges., Vol. 26).

SZAFER, W., 1935: The method of isopollens applied to the investigations of the history of trees by means of pollen analysis (Proc. Intern. Bot. Cong., Amsterdam 1935, Vol. II).

VOLKENS, G., 1912: Laubfall und Lauberneuerung in den Tropen (Berlin).

WETTSTEIN, R., 1892: Die fossile Flora der Höttinger Breccie (Denkschr. d. Akad. Wiss. Wien, Vol. 59).

WHITE, O. E., 1926: Geographical distribution and the cold-resisting character of certain herbaceous perennial and woody plant groups (Brooklyn Bot. Gard. Record, Vol. 15, No. 1, pp. 1–10; Brooklyn Bot. Gard. Contrib. No. 46).

WHITE, O. E., 1928: Mutation, adaptation to temperature differences and geographical distribution in plants (Brooklyn Bot. Gard. Contrib., No. 53).

WULFF, E. V., 1937: Essay at dividing the world into phytogeographical regions according to the numerical distribution of species (In Russian, English summary; Bull. Appl. Bot., Gen. and Plantbr., Ser. I, No. 2. First publ. as brochure in 1934, German summary).

Chapter X

HISTORICAL CAUSES FOR THE PRESENT STRUCTURE OF AREAS AND THE COMPOSITION OF FLORAS

From the preceding chapters it is clear that in many cases the structure of the areas of species and the composition of floras cannot be explained by existing factors. The present distribution of any given species is a reflection of the geological revolutions and climatic changes that have occurred on our globe during the entire period of existence of that species. An elucidation of these great changes in the surface of our planet is the task of historical geology and paleogeography, a task as yet far from fulfillment. Consequently, the elucidation of the history of areas, the most difficult task of biogeography, likewise falls far short of achievement. In the present chapter we can, therefore, do no more than examine the chief theories that have been advanced and point out which give the most plausible and satisfactory explanation of the knotty problems of historical plant geography.

From very ancient times—at first without adequate foundation and later, with the development of geology and biogeography, on the basis of numerous data—the conviction has been held that the distribution of lands and seas was not always the same as now. For, if it had been, a considerable number of facts, both of a geological and biogeographical nature, would be inexplicable. The sedimentary character of the rocks covering extensive territories on the continents and the finding in these rocks of fossil marine animals testify to the fact that at one time seas covered these parts of the continents. That many islands formerly constituted a part of the mainland is shown by the geological structure of these islands, by their fossil and extant fauna and flora, and by the finding of submerged trees in various straits and channels, *e.g.*, in the English Channel. Furthermore, the outermost edges of continents and islands do not necessarily coincide with their shore lines, as their outer margins often lie submerged under so-called "shelf seas". The latter differ in extent, but their boundaries may be ascertained with considerable precision. The determination of these boundaries gives certain clues to the changes that have taken place in the distribution of lands and seas on the globe and on the probable existence in former times of connections between bodies of land now separated by the sea. Biogeographical data, in many cases, indicate that these changes occurred at a comparatively recent date.

1. *Theory of Land Bridges.*—We have already seen that a considerable number of cases of discontinuous areas of plants (and these might be supplemented by an equal number of instances of similarly distributed animals) cannot be regarded as accidental and require explanation. Often there seems to be only one possible explanation, *viz.*, that there formerly existed some sort of connection between the isolated habitats and that the now discontinuous areas were formerly continuous. Hence, many investigators have assumed that at one time

great "lost continents" or land-bridges of one form or another connected the continents now separated by oceans. On the basis of geological and paleontological data for different geological periods there have been postulated various connections between the continents, presumably later having sunk to the bottom of the sea and having been replaced by the upheaval of other land-bridges connecting other bodies of land. According to this theory, the distribution of land and sea has undergone constant change during the long history of the earth.

The complex conformation of these putative land-bridges clearly testifies to the artificial character of the hypotheses resting upon their probable rôle. This, together with the fact that there are a number of geophysical and geological arguments against this theory, has made it necessary to seek for new ways of explaining the former continuity of the now-discontinuous areas of organisms, all the more since the time when these land-bridges were supposed to exist is not always such as to be able to account for such continuity.

Even as regards biogeography, to which the creation of this cumbrous theory was a concession, it far from solves all the incomprehensible moments in the distribution of organisms. In particular, it leaves unclarified why plants in former geological periods grew in regions outside the climatic zones in which their present habitats are found, and also, which is very important, it gives no satisfactory explanation of discontinuous areas. If identical or closely related species are found on two continents separated by an ocean, we cannot explain this discontinuity of area merely by assuming the former existence of an intervening continent where the ocean now lies. It would likewise be necessary to assume that over the entire extent of this great "lost continent" there existed like ecological conditions, similar to those in the outlying portions of the area of the given species, these portions having been preserved in the form of isolated fragments of a once-continuous area. This can hardly be regarded as an acceptable hypothesis.

2. *Theory of the Permanence of Oceans and Continents.*—Against the theory of land-bridges, despite its partial fulfillment of the requirements of paleontology and biogeography and its acceptance by many geologists, there were advanced, beginning with the middle of the past century, a number of serious objections. On the basis of these objections a new theory was proposed, the theory of the permanence of oceans and continents. Among biologists this theory found many supporters, the most outstanding being DARWIN and WALLACE. The arguments advanced against the existence in the past of great continents that subsequently subsided to the bottom of the sea are so weighty that biogeographers cannot fail to take them seriously into account. These arguments, in brief, are as follows:

Upon the upheaval of continental masses in the area of our present oceans the great mass of water that had been in these oceans would have inundated all the continents, both old and new, except for the highest mountain peaks. Thus, the very continental connections that it was desired to create by the assumption of the existence of such "lost continents" would not have existed.

But if it be assumed that the position of the oceans and continents in former periods of the earth's history was not the same as at present, that where there are now continents, or at least over part of their territory, there were deep oceans and where the latter now are there were continents, then in the rocks of which our continents are composed there should be, over considerable expanses of territory, deep-sea deposits, attesting that at one time the given continent was at the bottom of the ocean. Such deep-sea deposits, however, are not to be found on our continents. This serves as grounds for concluding that our present continents were never covered by oceans and, in contrast to the ocean depths, always constituted elevated land-masses. On their surface in former geological periods, just as at present, there existed relatively shallow seas, but a large part (at least one-third) of their surface was always dry land (SOERGEL, 1917, p. 11). Slight variations in the level of the ocean, amounting to as much as several hundred meters, accompanied by a depression or elevation of the strand lines, may change considerably the contours of our continents, but they cannot affect their permanence as a whole nor the permanence of the ocean beds.

To geophysical objections to the land-bridge theory there may be added objections based on geological and paleontological data. These bear witness to the absence of transitional forms between the fossil marine faunas of the earlier geological periods and the sudden appearance of whole groups of species not connected with those of the formations preceding them in sequence of time, as would be required by the theory of evolution. These facts indicate that the successive phases of development of the inhabitants of the sea did not take place on the surface of our continents but must have taken place within the boundaries of the oceans as at present constituted. Hence, the following conclusions are drawn by the advocates of the theory of permanence (SOERGEL, 1917, p. 15): —

1. The great areas now occupied by oceans must have been thus occupied always, at least ever since the Pre-Cambrian era.

2. Fossil marine faunas have no roots on our present continents; they merely represent the repeated inland migrations of sea animals. (This is further confirmed by the absence in the faunas of present-day inland seas of elements that have preserved features of a deep-sea origin).

3. The territory occupied by our present continents has constituted a habitat of marine fauna always in contrast to the territory of the present oceans, a circumstance which is only understandable if one assumes that the former was an elevated territory, subject to alternate upheaval and depression, *i.e.*, a territory of a continental character. This means that the continents must have been permanent.

These conclusions, however, leave entirely unexplained the biogeographical data that indicate the need of assuming that the continents were at one time connected. It is true that now even advocates of the theory of permanence try to find a way out of the contradictions created. They state that, though we cannot concede the existence of large trans-oceanic land-bridges, we may concede the existence of some sort of connection between the continents, *e.g.*, narrow land-bridges between North America and Europe, Australia and South America,

Australia and Asia, Madagascar and Africa, the Antarctic continent and South America (Soergel, 1917). But this concession to biogeography is not founded on any new facts. Moreover, conceding the existence of such narrow land-bridges does not, by any means, solve all the unexplained problems in the distribution of organisms; it is necessary, in order to explain the latter, to presume still other connections not compatible with the theory of permanence. From this vicious circle these early theories provide no exit. An exit is found, in our opinion, only in Wegener's theory of continental drift, which we shall later discuss in detail.

4. *Pendulum Theory.*—An attempt to reconcile the contradictions involved in the two theories above outlined was made by the advocates of the so-called "pendulum theory", that aimed also to explain, on the basis of climatic changes in former geological periods, those cases of plant distribution that are not at all in accord with present-day climatic zones.

Changes in the position of the climatic zones on the globe—attested to by data on the distribution of fossil plants, showing, for instance, that at one time there was a rich flora within the area of the present Arctic region—must presumably have been caused not by a different location of the sun in relation to the earth but by a different location of the continents in relation to the sun. The pendulum theory explains these phenomena by assuming that periodic changes have occurred in the position of the poles caused by their oscillating back and forth like a pendulum. This theory was first advanced by the geologist Reibisch and later elaborated on the basis of biogeographical data by Simroth (1914). These investigators start from the assumption that the earth, besides an axis of rotation and poles of rotation located at the north and south ends of this axis, has an axis on which it oscillates like a pendulum. The two poles of this latter axis—known, in Simroth's terminology, as "Schwingpolen"—are located one in Ecuador and the other in Sumatra. Not subject to the oscillatory motion are only this axis and its two poles, Ecuador and Sumatra. These points alone remain fixed and under constant tropical conditions, while all the other points on the earth's surface are subject to periodic changes in climatic conditions induced by the oscillatory motion. Moreover, the degree of these climatic changes is determined by the distance from the equator. The greater this distance, the greater the deviations in climate suffered by any given point on the earth's surface. Conversely, those regions of the earth nearest to the equator, particularly those nearest to the "Schwingpolen", are subject to the least changes, climatic and physiographic, and, consequently, in these regions ancient plant and animal forms have been preserved to a much greater extent than in those regions located on the outer arc of oscillation.

If the pendulum theory is accepted, there is no need for the theory of land-bridges, since changes in the level of the sea induced by this oscillation of the earth would suffice to cause the joining together and disjoining of different parts of the earth's land surface and, moreover, in such a way as serves to explain the distribution of plants and animals. Simroth's theory of the distribution of organisms is based on those changes in the surface of the continents, particularly climatic changes, induced by the shifting of the position of the continents as a result of this oscillatory motion.

Every animal and plant, in the process of multiplication, naturally tends to spread within the limits of the same climatic zone within which it arose. Consequently, in the absence of obstacles to such distribution, their areas should encircle the globe in a band covering the territory occupied by the given climatic zone. This regularity in the distribution of organisms is upset by the oscillation of the earth. Thus, a plant or animal finding itself in the region of the outer arc of oscillation will be mechanically evicted from the climatic conditions to which it has been accustomed and will be forced to migrate to the west or east in order to return to its normal habitat conditions. This serves as an explanation of those discontinuous areas consisting of two halves symmetrically located on opposite sides of the arc of oscillation ("symmetrische Punkte" in SIMROTH's terminology). Moreover, if a species, as a result of such a change in habitat, does not undergo, in the process of adaptation to the new habitat conditions, vital changes in its morphological structure but acquires only a few, slight modifications, we shall find at these two points vicarious species. As an example of such "symmetrical points", we may take Japan and California. In case of the distribution of organisms farther inland in western North America and eastern Asia within the limits of the same latitudinal zone, we may speak of the *horizontal symmetry* of the distribution. This type of symmetry occurs chiefly during the polar phase of the oscillation. During its equatorial phase, on the other hand, the habitats shift to a very hot climatic zone, compelling marine animals, in order to attain more temperate climatic conditions, to go farther down into the sea and to adapt themselves to deep-water conditions, while terrestrial animals must ascend the mountains or migrate to the north or south, *i.e.*, in a meridional direction. As a result, we have distribution characterized by *meridional symmetry*, which is of most frequent occurrence in the outer arc of oscillation.

SIMROTH based his theory chiefly on carefully elaborated data on the geographical distribution of animals, but he also gave a number of examples of plant distribution that likewise agreed with his theory. Such agreement is expressed, first of all, in the actual existence of symmetry in the present-day distribution of certain species and in the fact that there are data establishing the existence in former times of habitats of these same species on corresponding parallels in the outer arc of oscillation. We shall here cite the most characteristic examples of those presented by SIMROTH and other investigators.

Let us first take a few instances from gymnosperm distribution. Out of fifteen species of the genus *Gnetum* seven grow in equatorial America, *i.e.*, in the region of the western "Schwingpole", one in Africa, and one on islands of the Pacific, while the six remaining species are grouped around the eastern "Schwingpole" in the eastern part of the Indian Ocean. Consequently, there is no doubt that we have here horizontal symmetry.

In the genus *Pinus* of most interest is the section *Taeda*, which comprises sixteen species. One group of these species grows in America

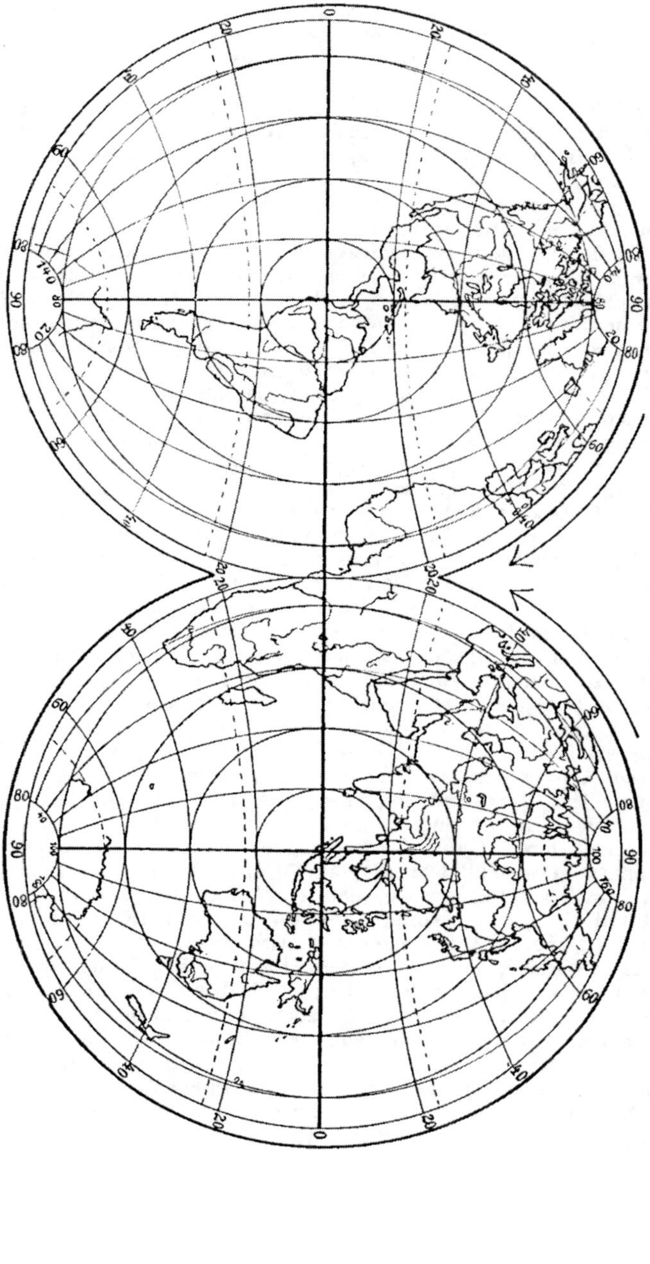

Fig. 20. — Map of the globe divided into two hemispheres along the arc of oscillation. In the center of each hemisphere is a "Schwingpole". The vertical meridian is the "culmination circle" (in the hemisphere at the left the 80th meridian, at the right, the 100th). The concentric circles represent the lines of motion of separate points during oscillation. (After Reibisch, from Simroth).

(in Florida and North and South Carolina, in the Rocky Mountains, in California, and in Mexico); the other group is found at a symmetrically opposite point in eastern Asia (the Himalayas, Tibet, and the Philippines). The area of one species, *P. canariensis*, growing in the mountains of the Canary Islands, seems to be an outlying spur of the former, more extensive range of this latter group.

Of two relic species of pine found on the Balkan peninsula, *Pinus Peuce* and *P. omorica*, the former is related to the eastern white pine, *P. Strobus*, of North America, and the latter to species of Manchuria and Japan. We have here, therefore, a clear case of the breaking up of a once-continuous area into two widely separated sections, one in the east and one in the west.

Analogous facts are likewise found in the distribution of angiosperms. For instance, the genus *Magnolia* has about 60 species, distributed in eastern Asia and in the Atlantic states of North America. *Liriodendron Tulipifera*, another member of the magnolia family, also grows both in China and in the Atlantic states of North America. The genus *Talauma*, also of the *Magnoliaceae*, has 32 species in India, Java, and the Philippines and 8 species in the New World (the West Indies, Mexico, Central America, and the northern part of South America). In a fossil state the *Magnoliaceae* are found in the outer arc of oscillation in deposits in Europe (beginning with the Cretaceous and ending with the Pliocene stage) and as far north as Greenland and Spitsbergen.

At first some biologists regarded the pendulum theory favorably, since they hoped to find in it a solution to those inexplicable features of the distribution of organisms about which we have already spoken. However, there are very serious objections to this theory which make it impossible to adopt it even as a working hypothesis. The chief objections are, first, that no cause can be found for such an oscillation, and, second, that geological data are not in accord with this theory. But even biologists advanced a number of serious objections, of which we shall note the most important. The assumption made by SIMROTH that Europe, as the region that has been most subjected to climatic changes, has constituted the chief center of origin of new forms is merely a hypothesis without any foundation. It is more logical to assume the exact contrary, since such great climatic changes as occurred in Europe during the Ice Age resulted in the creation of nothing essentially new. Europe, as compared with the other continents, is a comparatively small territory, and during the Tertiary period it was even smaller than now, so that it is difficult to believe that it constituted the place of origin of most of the flora and fauna of the world. Moreover, the existence of such centers of species-formation in other continents, *e.g.*, Asia, is beyond any doubt. Two other important objections are that many of the facts in the distribution of organisms cited by SIMROTH as caused by such oscillation of the earth may be explained, without assuming such oscillation, on the basis of ecological and edaphic data, and, lastly, that not all cases of discontinuous areas can be explained by the horizontal or meridional migration of species.

Nevertheless, the fact of symmetry in the discontinuous areas of many species and the fact that the break in these areas occurs precisely in the Euro-African sector were quite correctly established by

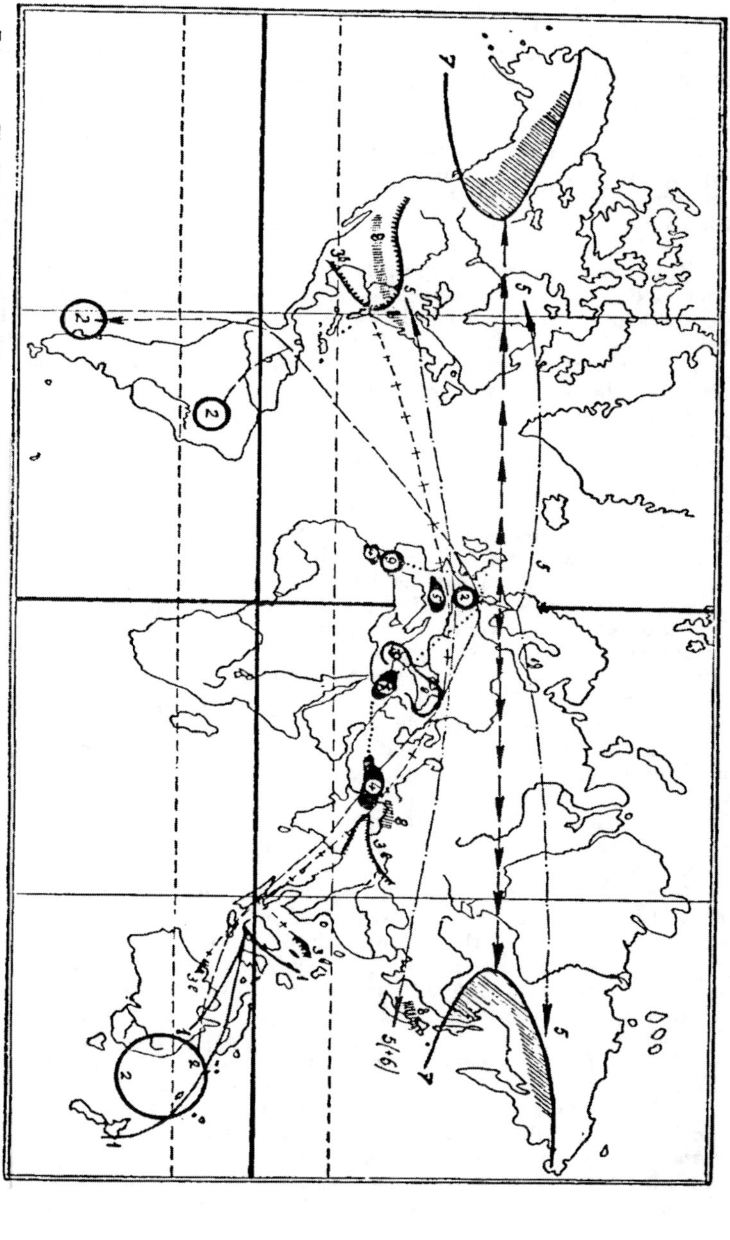

FIG. 21. — The Distribution of the *Coniferae*, according to SIMROTH: 1, *Dammara*; 2, *Araucaria*; 3, *Pinus Taeda*; 4, *Cedrus*; 5, *Larix*; 6, *Pseudolarix*; 7, *Picea sitchensis*; 8, *Tsuga*; 9, *Abies pinsapo*.

SIMROTH; this symmetry, however, may be ascribed to other causes, as we shall show below.

5. *Theory of the Polar Origin of Floras.*—Another attempt to avoid the need of trans-oceanic land-bridges to explain the present distribution of floras is found in the theory of the origin of the latter from a single center lying in the north polar region, whence they spread radially toward the south in three directions—through Europe into Africa, through Asia and Malaysia into Australia, which were at that time connected, and through North America into South America. As a basis for this theory there served the investigations of HEER (1868) of the fossil flora of the Arctic.

HEER, in a number of papers, later collected in his "Flora Arctica", pointed out that in former geological epochs there grew in the Arctic woody plants now found only in temperate or subtropical regions. His data showed that climatic conditions in the Arctic during the Tertiary period were entirely different from now and supposedly confirmed the view that the polar region was the initial center of origin of floras.

This point of view was first advanced by FORBES and was later developed by DARWIN, BERRY, and others. It is based on data indicating that beginning with the Tertiary period the floras and faunas succeed one another in such a way that each successive one forces its predecessor to the south. This shifting of floras and faunas presumably began from the moment of the differentiation into climatic zones, which, according to the views then held, did not take place until the end of the Cretaceous or the beginning of the Tertiary period, prior to which the climatic conditions of the earth were allegedly uniform. Consequently, the lowering of temperature conditions, which reached its apogee during the Ice Age, must have had its effect on the vegetation of the entire globe and must have induced a migration of floras from north to south. On the basis of the foregoing, the proponents of this theory held that the flora that had inhabited the body of land encircling the North Pole, which body of land prior to the Ice Age was presumably even larger and more compact than now, constituted the initial flora from which arose all the vegetation at present inhabiting the earth, and that the steady decrease in temperature in the polar regions forced the vegetation ever farther and farther south (FÜRSTENBERG, 1909).

In order to explain the similarities in the floras of South America, Australia, and South Africa, and the affirmations of HOOKER as to the circumpolar nature of the antarctic flora, it was presumed that there was a trans-equatorial migration of the flora of the northern hemisphere into the southern, although DARWIN and HOOKER had themselves suggested that in the past there might have existed an antarctic continent embracing what are at present separate islands and the extremities of the continents of the southern hemisphere. Subsequently, the existence of such a continent was generally accepted, the remains of fossil flora found within the limits of present-day Antarctica having confirmed not only that the lands of the southern hemisphere had formerly been connected with one another by way of this Antarctic continent but also that there had occurred a change of climatic conditions in the Antarctic just as in the Arctic, in the sense of a decrease in temperature. Hence, there was created a basis for the presumption that there probably had been at one time a center of species-formation in the region around the South Pole, similar to that assumed for the North Polar region, and a migration of floras, under the influence of climatic changes, from south to north. Thus, the mono-boreal theory of the origin of life was replaced by the theory that this rôle was played by the lands encircling both poles.

The theory of the polar origin of floras, in the light of our present knowledge, cannot be accepted. It has now been established that climatic zones have existed during the entire history of the earth and that ice ages occurred not only in the Quaternary period but also during other geological periods, the glaciated regions, moreover, not being located in the present polar areas but in other parts of the globe.

The assumption that there were only these polar centers of development of floras is likewise contraverted by the fact that the existence of other centers of species-formation has been definitely established. The view advanced by HALLIER (1912), GOLENKIN (1925), IRMSCHER (1922, 1929), and others that the tropics served as a center of origin of the angiosperms, whence they at various times penetrated into temperate regions, has much data to support it, provided it is accepted that there took place shiftings of the tropical zone.

An approach to the solution of all these enigmatic moments in the history of the earth, making it seemingly impossible to explain the past distribution of its floras, has been provided by the theory of continental drift.

6. *Theory of Continental Drift.*—From the foregoing we have seen that the two chief theories as to the past history of the earth's surface —the theory of the permanence of the oceans and continents and the theory of trans-oceanic land-bridges—are mutually contradictory. A solution of this riddle may be found, if one accepts the permanence not of the separate oceans and continents as such but the permanence of the relative area of land and sea taken as a whole. Then, by assuming, as WEGENER (1929) does in his theory of continental drift, the possibility of a horizontal drifting of the continents, which, so to say, float on an underlying viscous substratum (the so-called "sima", composed of basic, igneous basalts), we are enabled, without departing from the principle of the permanence of oceans and continents, to explain the existence of connections between the continents, not on the assumption that there formerly existed additional continents where there is now sea, but on the assumption that our present continents were formerly in direct contact with one another.

According to this hypothesis, it is assumed that as late as the end of the Paleozoic era the continents were all united in one great continent, Pangaea, which only in the Mesozoic era begins to rift apart, two meridian lines of rupture being formed. These lines of rupture between Euro-Africa and America, on the one hand, and between Africa and India, on the other, lead to the creation of the Atlantic and Indian oceans. The separation of South America from Africa becomes wider and wider during the Cretaceous period, but even at the beginning of the Tertiary period there still persists a slight connection between the northeastern coast of South America and the west-central coast of

Africa, the complete separation of these two continents taking place only after the Eocene stage. At the same time, the sea separating America from Africa becomes wider and wider, due to the drifting of America westward.

The connection between Africa and India through Madagascar likewise persisted as late as the beginning of the Tertiary period, being broken only in the Eocene stage due to the drifting of India northward. This movement of India northward is confirmed, according to SAHNI (1936), by paleobotanical data. The Permo-Carboniferous flora of southern Asia belongs to the type of *Gigantopteris*, *i.e.*, it is a tropical flora, while the flora of India of the same period is of the *Glossopteris* type, undoubtedly adapted to a temperate climate. With India and Asia located as at present the contiguity of these two widely different floras would be entirely inexplicable. It must be presumed that India at that time lay considerably farther south and was separated from Asia by the Tethys Sea. Only later did India drift northward, resulting in the seeming juncture of these two floras.

At an earlier period, *i.e.*, in the Jurassic, Australia broke away from India and Ceylon and Antarctica from Africa. The latter, still retaining connection with South America, drifted in a southeast direction. Sometime during the Tertiary period Australia separated from Antarctica, which, however, preserved until the Quaternary period its connection with South America. Not until the Quaternary period, due to the westward drift of the Americas, did Antarctica break away from South America and drift farther and farther in the direction of the South Pole. At about the same time, during the Ice Age, Greenland broke away both from North America and Europe, thus causing the isolation of these two continents from each other.

From the foregoing we may conclude that the contact between Europe and North America persisted until the Quaternary period, between Africa and South America until the Eocene stage, between Africa and India also until the Eocene stage, between Australia and India until the Jurassic period, between Australia and Antarctica and South America until the middle of the Tertiary period, and, lastly, between Antarctica and South America until the Quaternary period.

DU TOIT (1937) gave in his recent book on "Our Wandering Continents" very important geological proofs of WEGENER's theory, at the same time introducing some modifications. According to his viewpoint, the continents were originally represented by two great land bodies. The southern, Gondwanaland, embraced Brazil, Guiana, Uruguay, Africa, Arabia, Madagascar, India, western and central Australia, and Antarctica; the northern, Laurasia, included central and eastern Canada, Greenland, Scandinavia, Finland, Siberia, and northern China. These two great continental masses were separated by the Tethys Sea. Regressions of the latter led for a short time to a connection between Gondwana and Laurasia (between Africa and Europe; between Indo-China and Australia).

The breaking up of Gondwanaland, according to DU TOIT, took place not earlier than the Cretaceous or, perhaps, the Tertiary period. India began its movement to the northeast, over a distance of 1,500 km., at the beginning of the Cretaceous period. During the Cre-

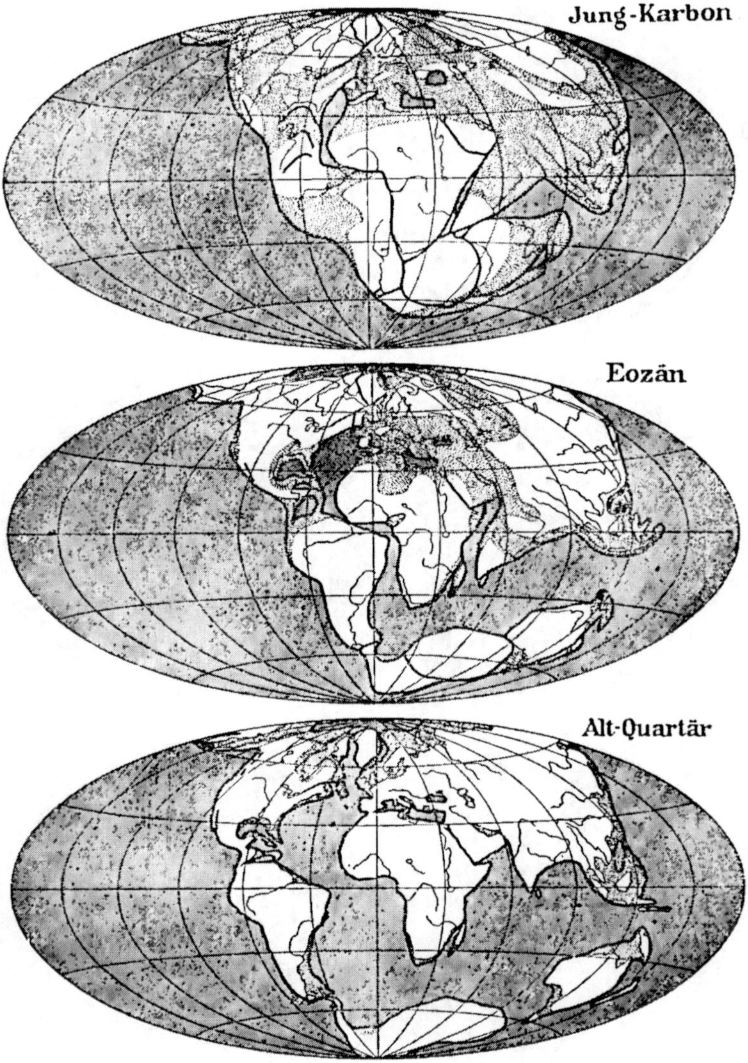

FIG. 22. — Reconstructions of the map of the globe according to data of the drift theory. Hatching = deep seas; stippling = shallow seas. Present-day rivers, contours of continents, etc. shown merely for purposes of orientation. (After WEGENER).

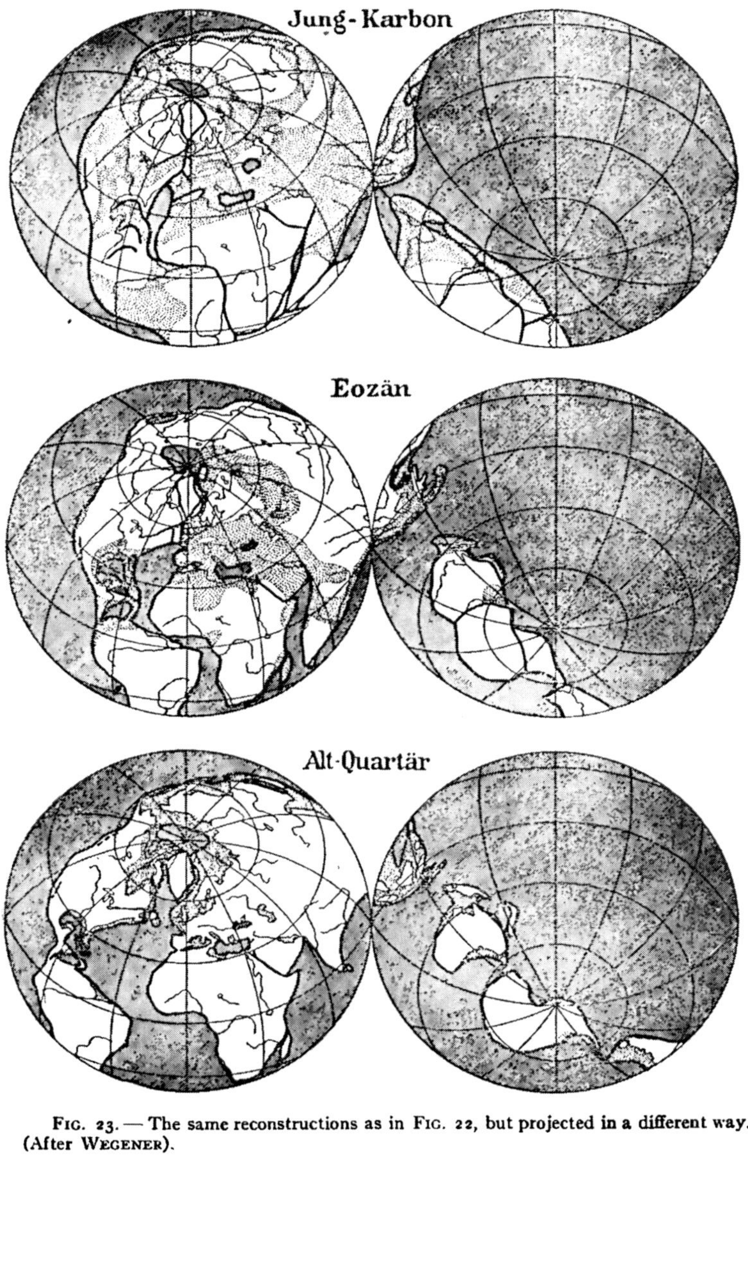

FIG. 23. — The same reconstructions as in FIG. 22, but projected in a different way. (After WEGENER).

taceous period New Guinea and New Zealand constituted the periphery of Australia. Not until the Tertiary period did they break away from Australia and drift off into the Pacific Ocean. The isolation of Australia, thus, presumably took place during the Tertiary period. At the beginning of the Cretaceous period it was still connected, by way of Madagascar and India, with southern Asia. The Andes of Antarctica constitute a continuation of the Andes of South America. The driftings of the continents induced changes in their position relative to the poles; this, in turn, led to changes in climatic zonation and, consequently, to changes in the distribution of living organisms.

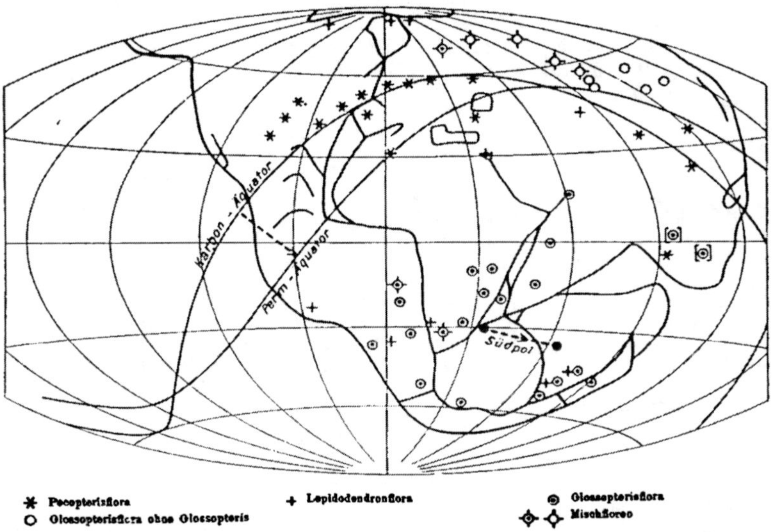

FIG. 24. — Distribution of floras and location of the equator during the Carboniferous and Permian periods. (After KÖPPEN and WEGENER).

On the basis of the foregoing assumptions riddles that formerly seemed insoluble in the past and present geographical distribution of organisms are solved. Among such riddles we may mention, first, the distribution of plants and even of entire floras in zones that, as regards their present climatic conditions, are not suitable to these plants, and, second, uniformity of vegetation, indicating that formerly there existed uniform climatic conditions in regions where at present there are marked differences in climatic and vegetation zones.

Another such riddle that long seemed insoluble is the absence in the Carboniferous period of periodicity in plant growth. In the case of our present-day plants this periodicity is expressed by the alternation of active periods (*i.e.*, periods of intensive growth and development) and dormant periods (during that part of the year when climatic conditions are unfavorable). One of the characteristic features of such periodicity is, for instance, the possession by deciduous and coniferous trees and shrubs of dormant buds, which remain closed during unfavorable periods of the year and renew development with the onset of favorable climatic conditions. In plants of the Carboniferous period, despite the existence of large trees, such as *Lepidodendron*, *Cordaites*, *Calamites*, etc., no such buds are found. As another even more characteristic feature of periodicity we may mention the annual rings in the stems of woody plants. In Carboniferous plants, distributed over a considerable part of the northern hemisphere and, to a less extent, in the southern hemisphere, no annual rings are found. This circumstance and also the absence of dormant buds testify to continuous and uniform growth, which could have taken place only under uniform climatic conditions, characterized by the absence of alternating seasons. Such climatic conditions correspond to those found at present in the equatorial zone with its tropical vegetation. But the plants of the Carboniferous

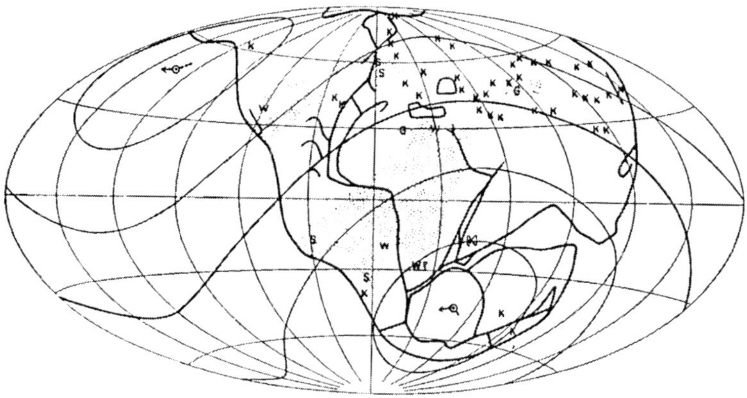

FIG. 25. — Distribution of swamps and deserts and location of the equator in the Jurassic period (K = coal, S = salt, G = gypsum, W = desert sandstone; stippling = arid regions. (After KÖPPEN and WEGENER).

period, though lacking periodicity, grew throughout the entire length and breadth of Europe.

In the flora of the Carboniferous period it is possible, according to POTONIÉ, to distinguish between elements of tropical origin, typified by the fern *Pecopteris*, and subtropical elements, typified by *Lepidodendron*, *Sigillaria*, etc. Whereas the former flora occupied in the Carboniferous period a quite limited, comparatively narrow strip, passing through North America and Europe and ending in eastern Asia, the subtropical (*Lepidodendron*) flora lay on both sides of this strip and had a considerably more extensive distribution. It is known as far north as Spitsbergen and as far south as the southern part of South America, occupying a latitudinal range of 120°.

Another riddle, which seemed no less difficult to solve, is the growing of trees near the poles as at present located. Moreover, as seems entirely incomprehensible, the long polar nights apparently had no effect on this vegetation.

Numerous attempts were made to solve these enigmatic peculiarities in the distribution and biology of fossil floras. It was assumed that plants of former periods possessed considerably greater ability, as compared with present-day plants, to adapt themselves to different climates and were considerably less sensitive to heat and cold. Or again, in order to explain the uniformity of climate over a considerable extent of the earth's surface, it was presumed that this uniformity of climate was due to the intense heat in the center of the earth and the insignificant amount of losses of this heat from irradiation owing to the fact that the earth was enveloped in a thick blanket of clouds. Lastly, it was considered possible to assume that the absence of annual rings was a peculiarity of plants of those times and that, consequently, this could not serve as a basis for conclusions regarding climatic conditions.

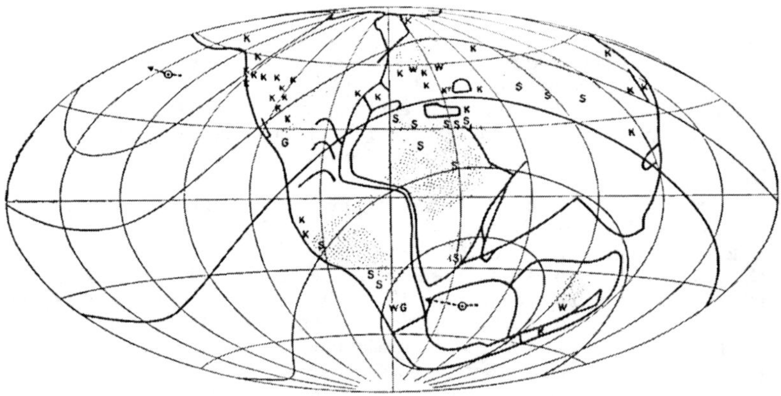

FIG. 26. — Same as in FIG. 25, but in the Cretaceous period. (After KÖPPEN and WEGENER).

These suppositions are refuted by the finding of traces of glaciers in the most ancient deposits of the earth and also by the fact that even in the Carboniferous period the climate was not everywhere uniform, since Carboniferous remains of trees having annual rings are known from the Falkland Islands and from Australia. In the floras of the succeeding geological stages periodicity in plant growth becomes of ever more widespread occurrence. All this indicates that climatic zones existed in past geological periods but that the location of these zones was undoubtedly entirely different from now.

It is likewise impossible to assume that there were any radical differences in the physiology and biology of plants of former geological periods. Paleobotanical data show that they were approximately the same as they are today. According to these data, fossil plants must have grown in climatic zones corresponding to the physiological peculiarities of these plants, whose requirements as regards light and heat must have been the same as those of their present-day descendants.

All the foregoing forces us to assume that the continents in past ages must have been differently situated with respect to the poles. A way out of all these difficulties in the geography of plants of former geological periods is provided, as we stated above, by WEGENER's theory of continental drift, in the light of which the climates of former geological periods receive an entirely different explanation. A detailed exposition of this viewpoint is given by KÖPPEN and WEGENER (1924) in a special work devoted to this problem. According to these investigators, in the Carboniferous period the equator passed through North America (from the southwest to the northeast), central Europe, the Caspian Sea, Asia, and the Sunda Isles. Central Europe was, therefore, included in the zone of equatorial rains, and it was, over much of its extent, submerged beneath the sea. Consequently, the Carboniferous deposits of central Europe must be of tropical origin; this is confirmed by the finding in such deposits of plants of the type of *Pecopteris*. Thus, the riddle of the finding in the Carboniferous deposits of central Europe of tropical flora lacking periodicity of growth is solved, and the assumption that the continents were formerly united explains the uniformity of climatic conditions over a considerable part of their territory. The theory of continental drift makes possible a new way of explaining problems involved in the paleogeography of plants, just as it aids in clearing up many formerly inexplicable facts in zoogeography.

The location of climatic zones in past geological periods was, according to WEGENER's theory, not at all the same as now; only gradually, beginning with the second half of the Tertiary period, did these zones begin to assume their present location. The shiftings of the climatic zones were accompanied by shiftings of floras, particularly on both sides of the Atlantic, in North America and Euro-Africa. Asia, on the other hand, suffered climatic changes to a considerably less degree, due to the fact that here the shiftings of the zones took place not symmetrically and in a circumpolar direction but asymmetrically. There are numerous facts both of a paleontological and of a floristic and faunistic character, long since noted by many investigators, that attest the relative constancy of the floras and faunas in the eastern part of the tropics of the Old World and, in particular, in the region of the Sunda Isles and eastern Asia. Thus, numerous Tertiary coal deposits have been found on the East-Asiatic coast and also on adjacent islands, *e.g.*, in the Soviet Far East (northern Sakhalin, Amur and Ussurian Regions), Manchuria, southern China, the Philippines, Java, Sumatra, and Borneo. According to data of Dutch geologists, there have been found on the Sunda Isles coal deposits from all stages of the Tertiary period. These facts, as well as the finding of species of palms in a fossil state, led KÖPPEN and WEGENER to the conclusion that throughout this region there has been a humid, tropical climate at least since the beginning of the Tertiary period.

IRMSCHER (1922, 1929), on the basis of ETTINGSHAUSEN's data, draws attention to the presence in the composition of the flora of this region not only of families but also of genera likewise found there in a fossil state. This has been confirmed by data of other paleobotanists, such as KUBART (1929) and KRÄUSEL (1929). MERRILL (1923), on the basis of studies of the flora of Malaysia, and KRYSHTOFOVICH (1933), on the basis of investigations of the flora of the Philippines, likewise arrived at the same conclusions. Even for regions as far north as northern Japan and Sakhalin we must conclude, on the basis both of paleobotanic data and of data of investigations of the present-day flora, that the composition of the vegetation has remained practically unchanged since Tertiary times, although here climatic variations were considerably more marked than farther south.

Turning now to the tropics of the New World, we do not find such constancy as regards climatic conditions; nevertheless, there were no such great climatic changes as took place in the more northern latitudes of North America and Europe.

The circumstances that we have just mentioned throw considerable light on the numerical composition of floras. Those regions of the earth that have been least subjected to climatic changes—southeastern

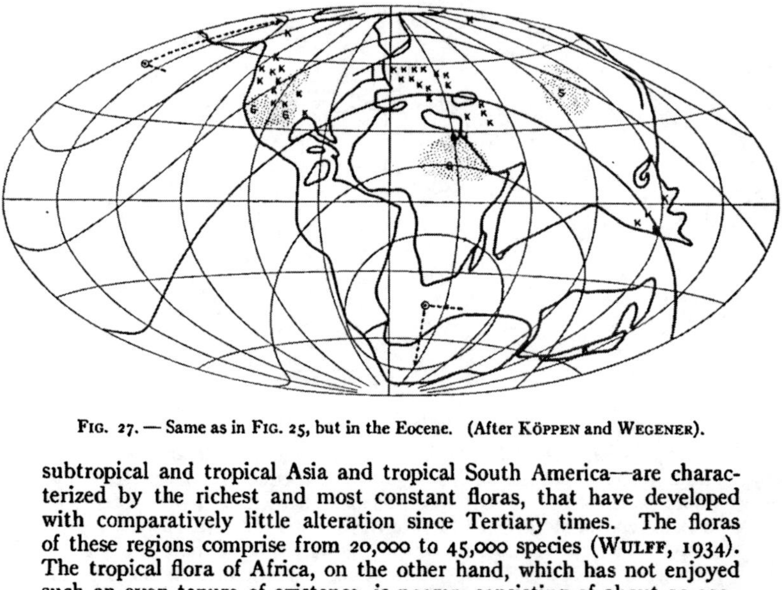

FIG. 27. — Same as in FIG. 25, but in the Eocene. (After KÖPPEN and WEGENER).

subtropical and tropical Asia and tropical South America—are characterized by the richest and most constant floras, that have developed with comparatively little alteration since Tertiary times. The floras of these regions comprise from 20,000 to 45,000 species (WULFF, 1934). The tropical flora of Africa, on the other hand, which has not enjoyed such an even tenure of existence, is poorer, consisting of about 10,000–15,000 species. We find a relative wealth of species in the Mediterranean floras and also in those of the southeastern and southwestern sections of the United States of America, which—though having undergone considerable impoverishment, expressed in the loss chiefly of hydrophytic species—have still preserved a number of ancient, Tertiary elements. This is particularly marked in the abovementioned sections of the United States and also in the southwestern part of the Iberian peninsula, the Istranja Dagh region in the Balkans, the northern coast region of Asia Minor, and western Transcaucasia, where, thanks to specific orographic conditions, these ancient elements found themselves under particularly favorable circumstances as regards protection from the decrease in temperature and humidity during the Quaternary period.

In all the rest of Europe and North America and also in northern Asia we find a great impoverishment of the floras, increasing from south to north, clearly testifying to the catastrophic changes which their floras have undergone since the Tertiary period.

From the foregoing it is quite clear that the origin of most genera of angiosperms, including tropical genera, was in all probability linked with those considerable territories embraced by the tropical and subtropical zones. Thus, in Europe the location of these zones in Cretaceous and Tertiary times made possible the origin and development of angiosperms throughout this entire continent, from its southernmost boundaries to our present arctic regions. Then the subsequent shifting of climatic zones, accompanied by changes in the habitat conditions of

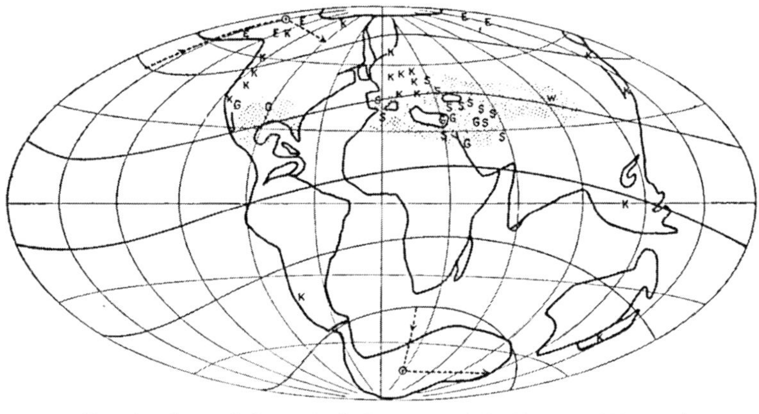

FIG. 28. — Same as in FIG. 25, but in the Miocene. (After KÖPPEN and WEGENER).

the vegetation, resulted in their dying out over considerable territories and in a shifting of the habitat regions. Consequently, our present-day tropics and subtropics and those regions characterized by a Mediterranean type of flora constitute territories where there has been preserved Tertiary vegetation, which formerly had a different and a considerably more extensive distribution. Hence there is no need to assume that the angiosperms originated in the polar regions, as do the advocates of the polar origin of floras, or that their origin was strictly confined to the region of our present-day tropics.

Phytogeographical Confirmation of the Theory of Continental Drift: —In the numerous phytogeographical papers published during the past two decades there are many data confirming the basic principles of the theory of continental drift. We can here present only a few of the most important of these data. The most comprehensive investigations, founded on an abundance of experimental data, aiming to test the theory of continental drift on the basis of the geographical distribution of plants, are those of IRMSCHER (1922, 1929). This investigator carried on very painstaking studies on the distribution of mosses and angiosperms, but before we take these up, we wish to set forth those regularities in the structure of areas which were established by IRMSCHER during the course of his investigations.

On the basis of the present and fossil distribution of the indicated group of plants IRMSCHER concluded that during past geological periods all the territory embraced by the torrid zone of those times was inhabited by a more or less homogeneous tropical vegetation. The north temperate and south temperate zones were likewise each occupied by a homogeneous vegetation, this greater homogeneity being due to the fact that the continents were differently located than now. Scattered fragments of this flora are still to be found on the continents of the southern hemisphere. The climatic changes that occurred must have affected similarly the areas of distribution of all groups of the plant kingdom. This uni-

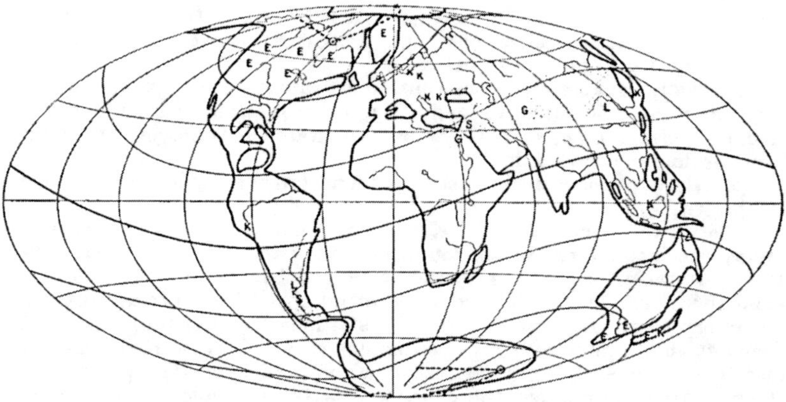

FIG. 29. — Same as in FIG. 25, but in the Pliocene. (After KÖPPEN and WEGENER).

formity of cause inducing uniform changes in the distribution of plants suffices to explain the regularities in the structure of areas first clearly established by IRMSCHER, though SIMROTH had noted them earlier in part.

The first of these regularities in the structure of areas is expressed in a certain *symmetry*, whereby along the meridians passing through Europe and Africa the given genera or species are entirely absent or are represented to only a very insignificant extent, as they often are likewise in Australia. Such breaks in an area may be explained only by cataclysmic or secular processes in former geological epochs that have destroyed its former continuity.

The second regularity is found in the *zonation* which characterizes the structure of discontinuous areas. Such zonation finds expression in the fact that the parts of a discontinuous area are located in a horizontal, not meridional line, so that we find them either in one of the temperate zones, or in the tropical zone, or in both temperate zones (bipolar areas).

The third regularity is found in the fact that the areas of many genera are *linked with definite continents*, areas embracing both America and Euro-Africa being rare, while there is a considerable number of areas embracing Euro-Africa, Asia, and Australasia. This linkage accords with the present location of the continents but, nevertheless, attests their different location in the past—a circumstance of great importance for further conclusions.

In order to understand the present area of any given angiosperm, it is essential to compare it with the same area supplemented by the territories where fossil remains of the genus or species or of closely related genera or species have been found. Such fossil remains occur chiefly in the northern hemisphere, particularly in North America and Europe, including their arctic regions. Most fossil forms have a distribution restricted to North America and Europe, temperate and subtropical species extending into the North Polar region. Tropical forms, on the other hand, are found chiefly in central and southern United States, in Europe, and in northern Africa. From the data derived from such comparisons of past and present areas we may draw the following very important conclusions:

1. Many genera with discontinuous areas, now having representatives only in America and Asia, in past geological periods had them also in Europe, so that the now separated portions of such areas were formerly connected.

2. Many forms were in past ages found farther north than they are at present.

The latter circumstance, *i.e.*, that the habitats of fossil forms of many species and genera are found in zones which at the present time are unsuitable, as regards temperature conditions, for their growth, combined with the established zonation of their distribution, makes it necessary to assume similar climatic conditions for the habitats of fossil remains that supplement a discontinuous area. If these fossil remains are found outside the climatic zone within which lies the given area, then, taking for granted the constancy of the physiological characteristics of plants, we must presume either that at that time the given climatic zone was broader or that, while retaining the same latitudinal range, it was subjected to a shifting from north to south. A solution to this problem we find in the fact that fossil remains have been found farther north than the habitats of their present-day representatives only in Europe and, to a lesser extent, in America, but they have never been found in Asia as far north as in the case of these other continents. On the other hand, the fossil remains from the tropics of Asia point to the fact that here climatic conditions have not suffered any great changes since the Tertiary period. Hence, it is necessary to conclude that the changes in climate did not affect equally all parts of any zone parallel to the present equator; thus, as regards our tropical zone, Asia remained untouched by these climatic changes. The only possible assumption capable of explaining this circumstance is that the climatic zones shifted obliquely, due to a change in the location of the poles.

On the basis of the foregoing it is further necessary to assume, in order to explain the former continuity of areas within any given zone, that the continents were at one time in contact with one another. Thus, if we take the southern hemisphere, we find a number of discontinuous areas, embracing South America and Australia, Africa and Australia, or South America, Africa, and Australia. The fact that these areas do not include any territory in the northern hemisphere cannot be regarded as accidental and indicates that these areas constitute parts of once-continuous areas. This connection between the continents, for instance, in the case of South America and Australia, must have been achieved by the shortest possible route, *i.e.*, by way of Antarctica. Hence we may conclude that the continents of the southern hemisphere were at one time connected, not only South

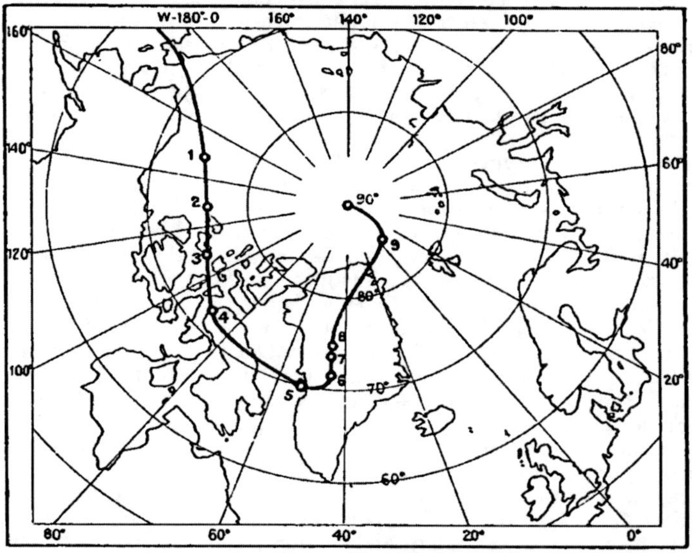

Fig. 30. — Shifting of the North Pole in its relation to Europe: 1, in the Miocene; 2–4, in the Pliocene; (4, Kansan); 5, during the Günz glaciation; 6, during the Mindel glaciation; 7, during the Riss glaciation; 8, during the Würm glaciation; 9, during the Baltic advance of the icesheet. (After Köppen and Wegener).

America, Australia, and New Zealand being in contact, but also South Africa.

On the basis of these fundamental features of the areas of angiosperms and the conclusions which we have drawn, we may now take up the problem of the genesis of these areas. From what has been said it follows that there existed a period in the history of many families of plants when their areas embraced very extensive territories, lying either within the boreal, north temperate, and tropical zones or within the south temperate zone. In the former case it is necessary to assume a direct connection between America, Euro-Africa, and Asia; in the latter case it must be assumed that it was possible for species to migrate directly from South America to Australia and also for an exchange of forms between Australia and Africa. This was a time when many families and genera had their maximum area. This period in the formation of areas IRMSCHER regards as the *first phase in their development.*

In the succeeding period these areas were subjected to changes under the pressure of various factors, of which changes in climatic conditions and in the location of the continents were the most important. These changes resulted in the areas becoming discontinuous, acquiring characteristics of symmetry and zonation. At the same time the genera occupying these areas continued to develop, there arose new species and genera, and as a result of their dispersal new boundaries for these areas were created. This constitutes the *second phase in the development of areas.* A characteristic feature of this phase is the clearly expressed adaptation of areas to the present location of the continents and their separation from one another. During this period the Atlantic Ocean was formed, separating Europe and Africa from America, and Australia and South America likewise became definitely separated.

Mosses.—Having made a study of discontinuities in the distribution of mosses, HERZOG (1926) came to the conclusion that their present distribution is comprehensible only in the light of WEGENER's theory of continental drift. IRMSCHER (1929) likewise studied the distribution of mosses from this point of view, and he arrived at the same conclusion as HERZOG. As a result of his investigations, IRMSCHER found that the distribution of mosses does not bear a chance, haphazard character, as would be the case if the transport of their spores by wind played as important a rôle as has usually been ascribed to it. His study of the distribution of mosses revealed just such discontinuous areas as are found in the case of conifers and angiosperms, this phenomenon being perhaps of even more frequent occurrence among mosses.

The most characteristic of such discontinuous areas are those whose separate parts are found in America, on the one hand, and in Euro-Africa, on the other. Such a discontinuity we find in the areas of two families (*Eustichiaceae* and *Oedipodiaceae*), each having only one genus, and in 36 genera, 8 subgenera and sections, and 144 species belonging to other families—a total of 190 taxonomic units. These areas lie, for the most part, in a single climatic zone; only a few embrace more than one zone. In some cases there is a gap in the area over the entire continent of Africa, while habitats are found on adjacent islands, as, for instance, in the case of the areas of the genera *Eustichia*, *Staberia*, and *Tortula*. The frequency of such very similar discontinuous areas cannot be accidental; it undoubtedly testifies to the fact that the now separate parts of these areas were at one time united. This lends support to the theory of continental drift, according to which America and Euro-Africa were formerly in contact.

That such a contact existed is likewise confirmed by those discontinuous areas whose separate parts lie not only in America and Euro-Africa but also in Asia and, in some cases, in Australasia. These areas are directly linked with those that show a break in the Euro-African sector. Thus, the genus *Rhegmatodon*, having a total of 12 species, is represented by 6 species in Central and South America,

4 species in Asia (from the Himalayas to Java), and only 2 species in the Euro-African sector and these only on islands (one on Fernando Po and one on Madagascar). The genus *Jaegerina* has disappeared altogether from the mainland and is now found only on islands, its distribution being similar to that of the genus *Ravenala*. One species grows on Jamaica, four on Madagascar, and one on Luzon Island in the Philippines. This paucity of stations within the Euro-African sector, observed likewise among the higher plants, may be explained only as a result of climatic changes that took place as a consequence of the shifting of the tropical zone far to the south, accompanied by an advance of the glaciers over most of Europe and a lowering of the temperature in central and northern Africa.

The impoverishment of vegetation in the Euro-African sector led, in

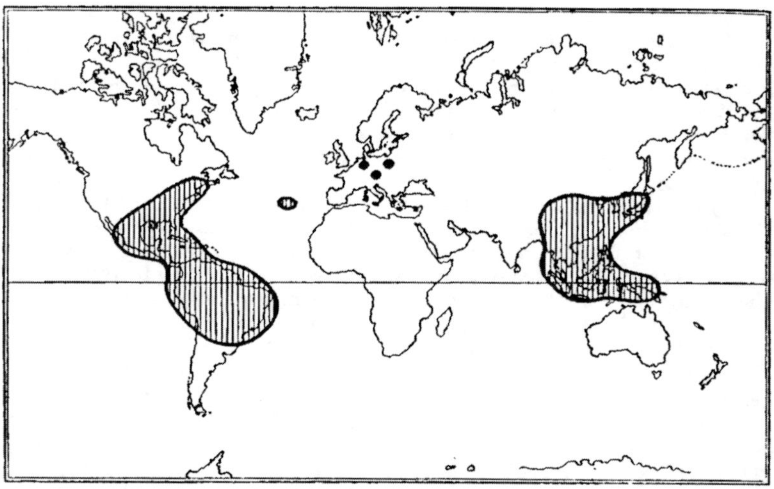

FIG. 31. — Area of distribution of *Clethra* (*Clethraceae*); Shading = present range; black dots = fossil occurrence. (After IRMSCHER).

the case of many genera, to a complete absence of representatives in Africa and the consequent formation of areas with two widely-separated parts, one in South America and one in Asia. The fact that genera with such areas comprise a very limited number of species indicates that they are very ancient. Some genera embrace a few species that have preserved their distribution in Europe but not in Africa. For instance, the genus *Hookeria* has one species, *H. acutifolia*, distributed in South America (Guadeloupe, Ecuador, Bolivia, Brazil) and Asia (the Himalayas, Ceylon, Java) and another species, *H. lucens*, distributed in North America and Europe. As offshoots of such areas we sometimes find (as also among the higher plants) spurs extending to Australia and Polynesia, which attests the former existence of a connection between Asia and Australasia.

There are similar discontinuous areas of mosses in the southern hemisphere, embracing territories in all or some of the following regions: Australia, New Zealand, and the southern extremities of Africa and South America. There are estimated to be 4 genera, 3 sections, and 18 species of mosses with areas embracing all the afore-mentioned regions, 2 genera and 16 species found only in South America and South Africa, 5 genera, 1 section, and 9 species in South Africa and Australasia, and 2 families, 14 genera, 8 subgenera and sections, and 64 species in South America and Australasia. The total number of taxonomic units having such southern discontinuous areas amounts to 142 (3 families, 33 genera, 12 subgenera and sections, and 94 species). It were just such discontinuous areas among other groups of plants that compelled HOOKER to believe that there must have existed a great Antarctic continent uniting in one land-mass the continents and islands of the southern hemisphere.

The fact that mosses, to an even greater extent than angiosperms, are characterized by discontinuous areas may be ascribed to their greater antiquity and greater ability to withstand changes in climatic conditions, which made it possible for them to survive the lower temperature at the close of the Tertiary and beginning of the Quaternary period in many places where the flowering plants were doomed to destruction.

Cycadaceae.—The origin of the *Cycadaceae* in the light of WEGENER's theory has been studied by KOCH (1925). The *Cycadaceae* are one of those ancient families whose origin must be referred to the Permian or Triassic period, *i.e.*, precisely to that time when, according to WEGENER, the rifting apart of the continents began. Consequently, a study of the history of their origin and distribution is of special interest.

The *Cycadaceae*, according to KOCH (pp. 68–69), arose directly from the Carboniferous flora in the equatorial zone of that time, *i.e.*, in central and eastern North America, central Europe, and the adjoining part of Asia lying at the same latitude, in so far as these latter regions were not covered by the great inland seas. During the succeeding geological periods the *Cycadaceae* spread to the north and south, embracing, on the one hand, extensive regions in northern Europe (at that time subtropical) and in Siberia, Spitzbergen, and Greenland and, on the other, the Mediterranean Region and north Africa. That they acquired such a wide distribution is attested by the numerous fossil occurrences of cycads, belonging to at least 34 genera and 278 species. This flourishing of the *Cycadaceae* reached its apogee at the beginning of the Cretaceous period, after which they begin to decline, conifers and angiosperms taking their place.

The history of the distribution of the *Cycadaceae* followed, according to KOCH (pp. 69–71), this course: One group of *Zamiaceae*, probably as early as the Triassic period, spread from central Europe to "Nordatlantis", whence it later extended its range southward as far as Central America and the northern part of South America. Of the four genera belonging to this group two, *Dioon* and *Ceratozamia*, eventually restricted their range to Mexico; the genus *Zamia*, however, is still represented by not less than 30 species in northern Brazil, Peru, Nicaragua (Granada Dep.), Guatemala, the Antilles, and Florida. The fourth genus, *Microcycas*, very closely related to *Zamia*, is now confined to Cuba. Another group of *Zamiaceae* spread southward from the equator of the Permian period into Africa, where, as the climate gradually became warmer, it reached, by way of eastern Africa, the Cape of Good Hope. Of this group there has been preserved the genus *Encephalartos*. From eastern Africa closely related forms spread, by way of Madagascar and the East Indies, to Australia, which lands at that time formed with Africa a vast, united land-mass (Gondwanaland). In east Australia (New South Wales) we find the genus *Macrozamia*, which is closely related to the African *Encephalartos* and the American *Zamia*.

The group of *Cycadaceae* proper, represented by the genus *Cycas*, is probably also of African origin. Although in Africa itself it has not been preserved, we find its most ancient representative, *Cycas Thouarsii*, on Madagascar and the Comoro Islands and a very close relative of this species, *Cycas circinalis*, in the East Indies. From these regions this group is presumed to have spread to adjoining Australia. Lastly, the East-Asiatic forms (*Cycas siamensis* and *C. revoluta*) reached their present habitats by way of the East Indies.

Schuster (1931) holds a different view. In his opinion the initial forms of the family *Cycadaceae* arose on the Eurasian continent. These Upper Triassic cycads (*Ur-Cycadaceae*) gave off two branches that spread from the ancient continent of Angara, one to Gondwanaland and one to North America. From the Tertiary descendants of these initial forms there arose our present-day cycads in different centers, such as the Himalayas, western Australia, South Africa, Mexico, and Central America. This investigator considers that his data controvert Wegener's theory, but, in order to explain them, he has to assume that there existed continents where now there is sea, which is refuted by modern geology.

Ginkgoales.—The distribution of the *Ginkgoales*—represented in our present-day flora by only one relic species, *Ginkgo biloba*, growing in Japan in forest reserves and possibly wild in China—has been studied by Shaparenko (1935, 1936) with the special aim of testing Wegener's theory. *Ginkgo* in past geological periods had a very extensive area in the northern hemisphere, as shown by the location of its fossil remains, known as *Ginkgo adiantoides*. Moreover, changes in the location of the area of *Ginkgo* in different geological periods coincide with remarkable exactitude with changes in the location of the 40th parallel as given by Köppen and Wegener (1924): —

> "The most ancient finds of this species — Ajakit [Siberia], Cape Boheman [Spitsbergen] — are located in the tertiary of Angarida, seemingly the center of the growth of this species. According to Wegener, this region should have lain at about 40° north latitude, which almost coincides with the latitude of the area of the living species now growing in a half-wild state" (Shaparenko, 1935, pp. 15–16).

In Cretaceous deposits *Gingko* is again found in the Far East, but it is also found in Greenland and in North America (Upper Cretaceous). In intermediate localities it has not yet been found. These fossil occurrences lend support to the presumption that these territories were at that time connected, since they cannot be explained by chance transport of seeds. The seeds of *Ginkgo*, as those of other gymnosperms of that period, cannot long remain submerged in sea water without injuring the embryo, so that transport by sea currents is improbable. Of birds at that time there existed only *Archaeopteryx*, and mammals had only just begun to make their appearance. There was also no possibility of wind transport, since *Ginkgo* has very large, heavy seeds without any adaptations for flight. This connection of Greenland and North America with Eurasia was broken, according to WEGENER, only in the Quaternary period. This genus extended its range eastward and westward from the indicated Angara center in strict conformity with the location of the 40th parallel. The absence at this time (at the close of the Jurassic and the beginning of the Cretaceous period) of any considerable shiftings of area in the meridional direction is in accord with the supposition of KÖPPEN and WEGENER that climate changed very little during these geological periods.

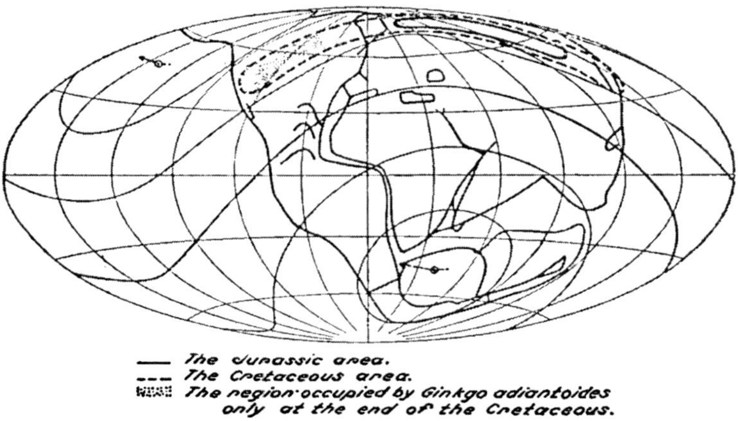

FIG. 32. — Areas of *Ginkgo adiantoides* Unger in different geologic periods. The geographic network (0°, 30°, 60°) and the shapes of the mainlands are those of the Cretaceous. (After SHAPARENKO).

At the beginning of the Tertiary period there took place an extension of the area of *Ginkgo* in North America toward the north, which corresponds to the change to a warmer climate that occurred here and reached its maximum during the Eocene stage. In the Oligocene and Miocene stages the climate over the entire Eurasian continent and also over North America began to get colder. This caused a shifting of the area of *Ginkgo* to the south. Prior to the Pliocene stage it had already reached as far south in North America as British Columbia and in Europe as Sterlitamak (Bashkiria). During the Pliocene stage *Ginkgo* became extinct in America, and in Europe there remained only a small, isolated area in southern France, which was completely wiped away by the Quaternary glaciation. *Ginkgo* has thus been preserved to the present day only in eastern Asia, a region which was not subjected to glaciation and which has suffered least of all from great climatic changes. This accounts for the preservation in the flora of this region of a number of ancient, relic elements.

Coniferae.—A test of WEGENER's theory on the basis of the distribution of conifers was made by STUDT (1926). The regularity established by IRMSCHER as regards the structure of areas was fully confirmed by STUDT in the case of the areas of conifers, a study of which led him to the following conclusions.

The first conifers made their appearance in the Upper Carboniferous and Permian periods. As regards their dispersal, we can, on the basis of paleobotanic data, establish three stages: (1) from the Upper Carboniferous to the Triassic period—dispersal throughout the north temperate zone of Europe and North America; (2) during the Mesozoic era there are formed two centers of distribution of conifers—in the southern hemisphere there are concentrated *Araucariaceae* and *Podocarpaceae* and in the northern *Pinaceae* and *Taxodiaccae* and later also

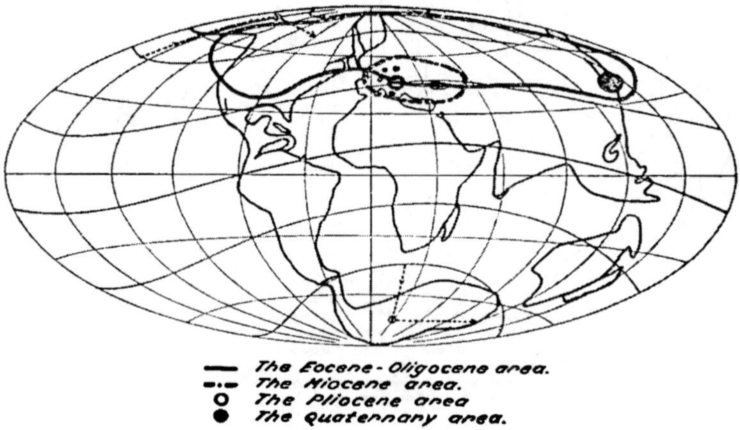

FIG. 33. — The Eocene-Oligocene, Miocene, Pliocene, and Quarternary areas of *Ginkgo adiantoides*. The geographic network and the shapes of the mainlands are those of the Miocene. (After SHAPARENKO).

Taxaceae and *Cupressaceae* (at this time Europe has an abundance of conifers, exceeding possibly in number even those of North America); (3) the Quaternary period is characterized by the predominance of conifers in North America and eastern Asia and the occurrence of isolated areas in the southern hemisphere. In Europe during the Ice Age a large part of the conifers died, while in North America and eastern Asia, thanks to more favorable conditions, particularly the absence of mountain chains running east and west that would have constituted barriers to the southward spread of vegetation, they were preserved in considerably greater numbers. In these latter regions in post-glacial times there also arose new forms, while Europe remained poor in species of conifers. Consequently, as IRMSCHER also established, Europe, which was characterized during the first and second stages by a great abundance and diversity of conifers, is during the third stage characterized by extreme poverty as regards conifers.

For a confirmation of WEGENER's theory it is exceptionally important to establish correspondence or lack of correspondence between the past distribution of plants and present-day climatic zones. As we have already noted, KÖPPEN and WEGENER believe that during the Cretaceous period and the beginning of the Tertiary period the tropical zone occupied a different position than it does at present. By glancing at the location of fossil occurrences of conifers, one is immediately struck by the fact that, besides the marked zonation of their distribution, they are found in the present-day arctic region in places located beyond the northern limit of distribution of woody plants and now lying under a permanent cover of snow and ice. Thus, according to data presented by KÖPPEN and WEGENER, on Spitsbergen there have

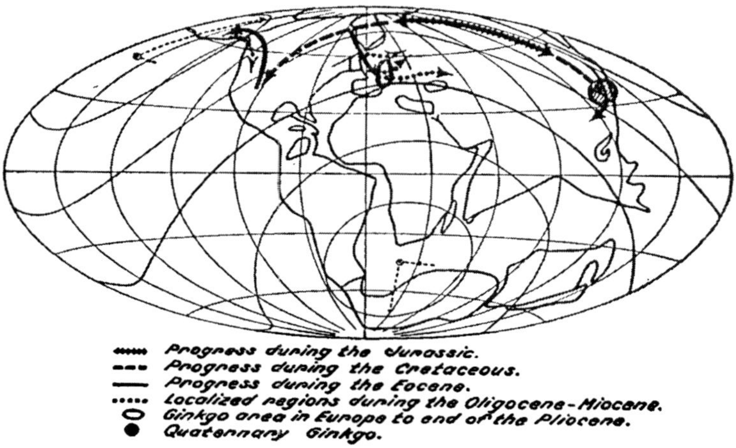

FIG. 34. — The distribution of *Ginkgo adiantoides* Unger in different geologic periods. The geographic network and the shapes of the mainlands are those of the Eocene. The distribution was effected by means of the connections between continents. The map shows the progress of distribution during the Jurassic, the Cretaceous, and the Eocene, the region in which the plant was localized during the Oligocene-Miocene, the ginkgo area which remained in Europe to the end of the Pliocene, and the Quarternary ginkgo area. (After SHAPARENKO).

been found fossil remains of 179 species of woody plants, conifers predominating; in the western part of Greenland, as far north as 70° N.—169 species; on the Island of Disko—282 species, of which 19 are ferns, 28 conifers, and 200 dicotyledons. Moreover, all these fossil woody plants are characterized by annual rings. In deposits of the end of the Tertiary period this type of vegetation in the arctic region begins to disappear and in the Quaternary period it vanishes altogether. At this time in the Antarctic there was still a subtropical climate, as is attested by the finding there, in Quaternary deposits, of *Araucaria Nathorstii;* the Falkland Islands were still heavily wooded and Seymour Island, now lying in a region of eternal ice, was also characterized by a subtropical flora at the close of the Tertiary period.

Further evidence of the occurrence of great climatic changes is provided by the asymmetry of areas, already mentioned as regards other types of vegetation and likewise characteristic of conifers. Thus, if we glance at the areas of conifers, we find: (1) uniformity in the composition of the coniferous floras of the Paleozoic and Mesozoic eras, often even to the extent of comprising the same species, due to the exceptionally wide distribution of many genera and species (*e.g.*, *Podocarpus* and *Araucaria*); (2) similarity between the Mesozoic conifers of North America (particularly those of the Pacific coast) and Europe (*e.g.*, *Sequoia*); (3) occurrence of such genera as *Araucaria*, now distributed only in the southern hemisphere, quite far north in the northern hemisphere; (4) zonal distribution of areas; (5) discontinuity of the present area of *Araucaria* with two widely-separated parts in South America and Australia.

By assuming, as WEGENER does, that in the Permian period all the continents formed one great land-mass (Pangaea) and that subsequently they broke apart and shifted their position in relation to the poles, all the peculiarities in the distribution of conifers become understandable.

Angiospermae.—The regularities in the structure of areas of angiosperms—established, as we have already noted, by IRMSCHER, on the basis of a study of the distribution of the various families belonging to this group of plants—led him to certain conclusions, which we shall now present in brief.

The facts established regarding the first stage in the development of the areas of angiosperms—fossil occurrences of tropical genera in Europe and North America, migration of plants from America to the east, the law of zonation—cannot be explained either by the theory of the permanence of the Atlantic Ocean or by the assumption that there existed a trans-Atlantic land-bridge from Greenland to Iceland and Scotland. In order to explain these facts, it is necessary to assume either that there existed a huge continent connecting America with Europe across the Atlantic or, which is the more probable, that these two continents were in direct contact with each other.

The second stage in the development of areas is very clearly demarcated from the first stage. If one were to assume the permanence of the oceans, it would be impossible to explain the entirely different configuration of areas in their first and second stages.

The picture drawn by WEGENER of the position of the continents before and after the breaking up of Pangaea could not be in closer accord with these first and second stages in the development of the areas of angiosperms. WEGENER's theory of continental drift not only makes plausible a close connection between North America and Europe but also provides a solution of the hitherto inexplicable riddle of the distribution of vegetation in the southern hemisphere. The fact that the floras of South America, Australasia, Africa, and the subantarctic islands have much in common, quite incomprehensible in the light of their present widely-separated locations, is at once elucidated by presuming that these territories in the Mesozoic era were united, thus making possible an interchange of vegetation.

IRMSCHER's final conclusions may be summarized as follows: (1) In order to understand the geographical distribution of plants, it is necessary to assume a shifting of the poles, and, as a corollary to this proposition, a shifting of the climatic zones; (2) of all the hypotheses advanced regarding the origin of our continents WEGENER's theory is most closely in accord with data on the distribution of angiosperms, and, hence, these data serve as a proof of the theory.

WEGENER's theory is likewise substantiated by the data presented by LAM in two papers (1932, 1934) devoted, respectively, to the floras of the Malay Archipelago and New Guinea. In the former, which constitutes a detailed, monographic study of the *Burseraceae* and of the phylogenetic relationships between its component species, LAM (1932, p. 300) comes to the conclusion that the primitive *Burseraceae* must have originated in South America, in Brazil, whence, thanks to a connection with West Africa and Eurasia that existed as late as the Upper Cretaceous or even as the Lower Eocene stage, species of this family reached eastern Asia and became established there, particularly on the islands of the Malay Archipelago. From the point of view of this investigator, the history of the *Burseraceae* and of their distribution agree quite closely with WEGENER's theory.

Of very great significance in this connection is SHAPARENKO's paper (1937) devoted to the past and present distribution of the tulip tree, *Liriodendron*. This investigator had as his special aim to test WEGENER's theory on the basis of data on angiosperm distribution. His work is of all the more value, since it is based not only on published data but on a first-hand study of a number of collections of the fossil remains of this plant. *Liriodendron* is represented at the present time by only two species: the first, *L. Tulipifera*, is distributed throughout the Atlantic section of the United States, from Florida to Michigan (between the 30th and 45th parallels); the second, *L. chinense*, has a more restricted area at about the same latitude but in eastern Asia, not far from the Pacific Coast, in the Kiang-si Province of China.

The first fossil leaves of the tulip tree are found in Cretaceous deposits in North America. The Cretaceous area of *Liriodendron* was located in the southeastern part of North America, the center of the area being at about the 30th parallel, according to the climatic zonation given for this period by KÖPPEN and WEGENER. The eastern end of the area lies a little south of this parallel and the western end a little north of it. This is in accord with the present lack of full coincidence between the isotherm and this parallel.

By the beginning of the Upper Cretaceous *Liriodendron* attains its greatest development in North America, but by the end of the Cretaceous period it has already become much poorer in number of species, and from the Tertiary period no fossil remains at all have been found. We have no data for judging as to whether this genus died out completely in America or whether, in connection with the onset of a much warmer climate, which attained its maximum in America during the Eocene stage, its area was shifted considerably farther north. The latter supposition seems to us the more probable, despite the fact that fossil remains of the tulip tree have not yet been found in Tertiary deposits in America.

But at this very time *Liriodendron* makes its appearance in Europe, which indicates that there existed a land connection between this continent and America. The earliest Tertiary remains of this plant, which have been referred to the Mexican species, *L. Procaccini*, are those from the Eocene stage found in England and Iceland. These fossil occur-

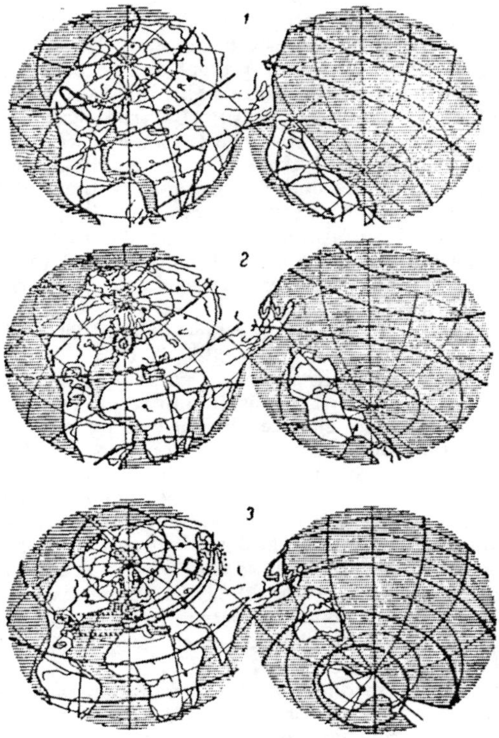

FIG. 35. — Area of the genus *Liriodendron* in different geologic periods: 1, Cretaceous; 2, Eocene; 3, Miocene, but areas during several geologic epochs are plotted: bent oval = Miocene; squares = Pliocene; dotted line = Quarternary. The map of the Cretaceous is taken from WEGENER, the others have been compiled according to WEGENER's data. The geographic network (0°, 30°, 60°) of the respective geologic epochs has been plotted on all three maps according to WEGENER's data. (After SHAPARENKO).

rences lie just south of the 30th parallel of those times, which is at approximately the same latitude as the center of the area of this genus during the Cretaceous period.

During the Miocene stage *Liriodendron*, represented in all probability by a considerable number of species, greatly extends its range, which assumes the form of a long, narrow strip, following the 30th parallel of WEGENER's reconstruction across the entire breadth of the Eurasian continent from western Europe to Japan. As it approaches the Pacific Coast, the area lies somewhat to the north of this 30th parallel. The change to an ever colder climate, that was initiated toward the end of the Tertiary period, led to the dying out of the tulip tree over a large part of its area. By the Pliocene stage we already have two widely-separated areas, one in southeastern Asia and the other in the southern part of western Europe. These two areas served as centers of origin of the two present-day vicarious species. The Asiatic area gave rise to *L. chinense* while the European area, according to paleobotanic data, was presumably the place of origin of *L. Tulipifera*. There is one circumstance that tends to cast doubt on this explanation of the origin of these species, *viz.*, that the latest fossil remains of leaves of *L. Procaccini* in Europe are those of the beginning of the Quaternary period, found in France. After this *Liriodendron* entirely disappears from Europe. On the Atlantic Coast of North America, on the other hand, after a prolonged lapse (during the whole of the Tertiary period), there are again found Quaternary fossil remains of the species *L. Procaccini*.

In explanation of these puzzling facts several suppositions may be advanced. Of these the most plausible is that during the Tertiary period in North America the area of this genus was shifted considerably to the north, where it persisted until the end of the Tertiary period, when, due to the climate then becoming much colder, it migrated back to the south to its former location. Although fossil remains of *Liriodendron* have not yet been found in Tertiary deposits of North America, it seems difficult to assume that this genus completely disappeared during this period from the flora of America. It is to be hoped that future paleobotanic findings will solve this enigma.

According to data of I. G. KNORING of the Botanical Institute of the Academy of Sciences of the U.S.S.R. (unpublished paper), there exists similar close correspondence between changes in the area of the genus *Liquidambar* and the changes in climatic zonation hypothesized by KÖPPEN and WEGENER.

In addition to these data on the distribution of separate families and genera, analyses of entire floras and studies of the history of their development have provided numerous instances showing WEGENER's theory of continental drift to be in full accord with the data on the migrations and interrelations of these floras in past geological periods. The material of such studies is too voluminous to be presented here even in summary form. It will be treated at considerable length in the author's forthcoming book, "A History of Floras" (to be published by the Academy of Sciences of the U.S.S.R.).

Phytogeographical Arguments Against the Theory of Continental Drift: — The theory of continental drift has been subjected to criticism on the part of some paleobotanists and phytogeographers. But the arguments advanced against it are comparatively few in number. Among the most important are those expressed by DIELS (1928, 1934, 1936), which may be summarized in the following three propositions:

1. The flora of the Atlantic Coast of North America is linked not with the flora of Europe but with that of eastern Asia;

2. The relic flora of Europe is related to the Asiatic flora and does not bear any American features;

3. The autochthonous flora of Australia bears a paleotropic character.

On the basis of these three propositions DIELS considers that WEGENER's hypothesis is incompatible with the data of botanical geography. DIELS' arguments were replied to at length in a paper by IRMSCHER (1929), to which we refer the reader for a detailed criticism. We wish here only to state that DIELS' categorical conclusion is, in our opinion, not convincing, since the relatively meager data advanced by him are refuted by counter-arguments, the number of which is constantly increasing. Thus, recent analyses of the elements of the flora of Australasia (SCHWARZ, 1928; LAM, 1934) have elucidated the relationships between these elements and have constituted a basis for conclusions quite at variance with those of DIELS. Moreover, the latter's arguments do not by any means cover all the phytogeographical problems that are solved by the theory of continental drift. In answer to his criticism it is possible to present a considerably greater number of biogeographical facts that are not contradictory to but confirmatory of WEGENER's theory. Lastly, DIELS' arguments may be answered in his own words: „Eine Theorie der Erdgestaltung soll uns nicht ein beliebiges Areal erklären, sie muss die grossen Züge der Florenbildung verständlich machen" (1928, p. 55).

Considerably more serious are the arguments against WEGENER's theory based on problems regarding the Pacific Ocean, in which field, from the point of view of biogeography, this theory requires certain modifications. According to WEGENER, the Pacific Ocean has existed during the entire history of the earth, and its islands constitute land fragments left at the time of the formation of this ocean in pre-geologic times. Assuming such an origin of these islands, it follows that all their fauna and flora are adventive in character. However, there are numerous biogeographical facts that refute such a conclusion. Consequently, many investigators of the organic life, particularly plant life, of the Pacific islands—*e.g.*, BECCARI (1886), HOLDHAUS (1909), HALLIER (1912), SKOTTSBERG (1925, 1928), VON IHERING (1927), GUILLAUMIN (1928), and CAMPBELL (1928)—consider that there must have existed trans-Pacific connections between America and Asia. These views have also found support on the part of a number of geologists (HAUG, 1912; GREGORY, 1930).

A similar view is advanced by DU RIETZ (1931) in his interesting paper on two new species of *Euphrasia* from the Philippines, *E. philippinensis* and *E. Merrillii*. The genus *Euphrasia* has two widely-separated centers of concentration of species, one in the northern hemisphere, in Europe, Asia, and the northern part of North America, and the other in the southern hemisphere, in Australia and the archipelago of islands to the north, New Zealand, and the southern part of South America. Between the groups of species located in North and South America, over the entire length of the two American continents, there are no connecting links. The South American species, however, possess a number of characters indicating that they are undoubtedly closely related to the species of Australasia, a circumstance undoubtedly attesting a former trans-Antarctic connection between these two ancestral centers. The two new species from the Philippines are very close in their morphological characters to the Japanese species, particularly those from Formosa. The latter, in turn, are closely related to *E. borneensis* (Mt. Kinabalu, Borneo), but the characters which they have in common are not shared by the Philippine species. *E. borneensis*, in its turn, shows close relationship with species growing in the mountains of New Guinea, where the genus *Euphrasia* became considerably more differentiated than in Borneo or the Philippines.

"The tropical species of *Euphrasia* occurring on the high mountains of New Guinea, Borneo, Luzon, and Formosa thus obviously form a natural connection and a gradual transition between the austral and the boreal parts of the genus" (Du Rietz, 1931, p. 529).

The data assembled by Du Rietz led him to the assumption that "the tropical *Euphrasia*-population . . . is not only the result of accidental colonization of isolated tropical mountains by species from the main distribution areas of the genus in the north and in the south, and the subsequent transformation of these species into the present tropical species, but that these tropical species are really remains of the old bridge once forming a more continuous connection between the boreal and the austral populations of the genus. In *Euphrasia* this trans-Malayan bridge . . . has obviously been the only connection between the boreal and the austral parts of the genus. The present distribution of the genus *Euphrasia*, together with the entire lack of traces of a direct relationship between the widely separated North American and South American *Euphrasia*-populations, clearly shows that the Andean trans-tropical bridge, so important in many other bipolar genera . . . has never been in use in this genus. On the other hand, there is a very distinct relationship between the South American population and the populations of Tasmania and New Zealand" (*Ibid.*, pp. 531–532). This last-mentioned circumstance clearly indicates that in the past there existed a trans-Antarctic connection between South America and Australasia.

Of exceptional interest is the discovery by Skottsberg in the subalpine belt of Juan Fernández of a single species, *E. formosissima*, endemic to this island. Most remarkable is the fact that this species shows close affinity not to the South American species but to the Australasian species, particularly those of New Zealand. Analogous relationships are found as regards the Juan Fernández species of several other genera, *e.g.*, *Halorrhagis*, *Ranunculus*, *Coprosma*, and *Santalum*, which indicates the former existence of a direct connection between this island and the islands of the western part of the Pacific. Thus, we seem to have here fragments of a former trans-Pacific land-bridge.

Du Rietz (*loc. cit.*, p. 538) considers that Wegener's theory with "its absurd phytogeographical consequences" cannot provide an explanation of the foregoing facts. Despite the interest and importance of the data presented by Du Rietz, his conclusions are too categorical. His data show the need of further research on the theory of continental drift, but touching, as they do, only one of its propositions, they cannot serve as grounds for discarding the theory as a whole.

As an authoritative reply to DU RIETZ we may cite the words of SEWARD (1934), who himself is by no means inclined to accept WEGENER's theory unconditionally. As a result of a study of fossil woody plants of the Kerguelen Archipelago, he came to the conclusion that: —

"We are not exclusively concerned with possibilities of plant-dispersal and the stocking of areas that are now oceanic islands; there is also a climatic problem. Phytogeography is not only intimately connected with geology; it may receive substantial assistance from geophysics. Improved methods of soundings have shown that some parts of the ocean-floor are counterparts of mountainous regions of continents rather than submarine plains; former isthmian links from continent to continent . . . may be imagined with reasonable justification, but it is very doubtful, if they in themselves will supply adequate explanations of climatic conditions necessitated by extinct floras. Vertical movements alone are insufficient; lateral displacement of continents seems to be an almost necessary assumption. It is stated that recent soundings reveal contours on the sea-floor comparable with those of continents; and that the South Antillean arc, geologically and geographically, shows a mixture of oceanic and continental characters. These and other arguments have been advanced against the acceptance of WEGENER's theory with its 'absurd phytogeographical consequences', which DU RIETZ (1931) considers are demonstrated by DIELS and other authors whom he quotes. Though it may be conceded that the difficulties presented by the past and present distribution of plants cannot be satisfactorily disposed of by the acceptance of WEGENER's theory as he propounded it, this concession does not preclude belief in some form of continental drift.

"The Wegener hypothesis, despite the serious criticism which it has raised, appeals strongly to the imagination. . . . It is by no means improbable that solutions of some of the many problems of Plant Geography—both past and present—will be found, not in the raising of foundered continents, but through the acceptance of the mobility of the earth's crust, as a factor not merely imagined but substantiated by evidence which, it may be suggested, will eventually be provided" (p. 736).

All these considerations show that by far not all the puzzling problems in the distribution of plants on the islands of the Pacific Ocean and the countries bordering on this ocean may be considered solved. Undoubtedly there also existed continental connections in the southern part of the Pacific similar to the connection between America and Asia in the region of the Bering Straits. Nevertheless, it is impossible to assume the former existence either of a lost continent occupying the basin of the Pacific Ocean or of a direct contact between America and Asia, as in the case of America and Euro-Africa. The history of the Pacific Ocean is entirely different from that of the Atlantic.

Although we have not enumerated here all the points in WEGENER's theory that are not in accord with biogeographical data, we shall not pursue this further. In our opinion they do not suffice to discard WEGENER's theory as incorrect but only indicate that in this theory there is still much that requires further elaboration and modification. In any event its significance for biogeography is beyond doubt. It has thrown light on a number of hitherto entirely incomprehensible moments in the geography of plants. It has shown us new ways out from that blind alley into which we were led by the mutually exclusive theories of land-bridges and of the permanence of oceans and continents. Even though, as KUBART (1926) pointed out, continental drift is not the sole factor determining the present areas of relic species, and even though WEGENER's theory still requires additions and corrections, it, nevertheless, constitutes the only plausible working hypothesis upon which the historical plant geographer may base his conclusions.

References:

Beccari, O., 1886: Malesia, Vol. III (Genoa).

Campbell, D. H., 1928: The Australian element in the Hawaiian flora (Proc. Third Pan-Pacific Sci. Congr., Tokyo 1926, Vol. I; Tokyo).

Diels, L., 1928: Kontinentalverschiebung und Pflanzengeographie (Ber. d. Deutsch. Bot. Ges., Vol. 46, pp. 49–58).

Diels, L., 1934: Die Flora Australiens und Wegener's Verschiebungs-Theorie (Sitzungsber. d. Preuss. Akad. Wiss., Nos. 31–39).

Diels, L., 1936: The genetic phytogeography of the Southwestern Pacific Area, with particular reference to Australia (*In* "Essays in Geobotany in honor of William Albert Setchell"; ed. by T. H. Goodspeed, Berkeley, Calif.).

Diels, L., 1938: (*See* Lam, 1934).

Du Rietz, G. E., 1931: Two new species of *Euphrasia* from the Philippines and their phytogeographical significance (Svensk Bot. Tidskr., Vol. 25, No. 4, pp. 500–542).

du Toit, A. L., 1937: Our Wandering Continents (London).

Engler, A., 1879: Entwicklungsgeschichte der Pflanzenwelt seit der Tertiärperiode, Vol. I (Leipzig).

Fürstenberg v. Fürstenberg, A., 1909: Die Polarregionen im Lichte geologischer und literarischer Forschung (Naturw. Wochenschr., Vol. 24).

Golenkin, M. I., 1927: Victors in the Struggle for Existence (In Russian; Moscow).

Gothan, W., 1925: Review of Koch's "Über die rezente und fossile Verbreitung der Koniferen" (Zeitschr. f. Bot., Vol. 17).

Gregory, J. W., 1930: The geological history of the Pacific Ocean (Nature, Vol. 125, No. 3159, pp. 750–751).

Guillaumin, M. A., 1928: Les régions floristiques du Pacifique d'après leurs endémismes et la répartition de quelques plantes phanérogames (Proc. Third Pan-Pacific Sci. Cong., Vol. I, pp. 920–938).

Hallier, H., 1912: Über frühere Landbrücken u. Völkerwanderungen zwischen Australien und Amerika (Meded. 's Rijks Herbarium, Leiden, Nos. 8–14).

Hanausek, T., 1908: *Wulfenia* und die Pendulationstheorie (Oest. Bot. Zeitschr., Vol. 58).

Haug, E., 1912: Traité de Géologie, Vol. 1 (Paris).

Heer, O., 1868–83: Flora fossilis Arctica, Vols. I–VII (Zürich).

Herzog, Th., 1926: Geographie der Moose (Jena).

Hill, A. W., 1929: Antarctica and problems in geographical distribution (Proc. Inter. Cong. Plant Sciences, Ithaca 1926, Vol. II, pp. 1477–1486).

Hill, A. W., 1930: Present-day problems in taxonomic and economic botany (Nature, No. 3178).

Holdhaus, 1909: Zur Kritik von Simroth's Pendulationstheorie (Verh. Zool.-Bot. Ges., Vienna, Vol. 59).

Ihering, H. von, 1927: Geschichte des Atlantischen Ozeans (Jena).

Irmscher, E., 1922, 1929: Pflanzenverbreitung und Entwicklung der Kontinente, I and II (Mitt. d. Inst. f. Allg. Bot. in Hamburg, Vols. 5 and 8).

Kerner v. Marilaun, F., 1930: Paläoklimatologie (Berlin).

Koch, F., 1924: Über die rezente und fossile Verbreitung d. Koniferen (Mitteil. d. Deutsch. Dendr. Ges.; Nachträge, *Ibid.*, 1925).

Koch, F., 1925: Die Cycadeen im Lichte der Wegenerschen Kontinent- und Polwanderungstheorie (Mitteil. d. Deutsch. Dendr. Ges.).

Koch, F., 1927: Zur Frage der fossilen und rezenten Verbreitung d. Koniferen (Mitteil. d. Deutsch. Dendr. Ges.).

Koch, F., 1931: Die Entwicklung und Verbreitung der Kontinente und ihrer pflanzlichen und tierischen Bewohner (Braunschweig).

Koch, F., 1933: Die Bedeutung der Wegenerschen Theorie für die Dendrologie (Mitt. d. Deutsch. Dendr. Ges., No. 45).

Köppen, W., 1931: Grundriss der Klimakunde (Berlin).

Köppen, W. and Wegener, A., 1924: Die Klimate der geologischen Vorzeit (Berlin).

Kräusel, R., 1921: Paläobotanische Notizen, IV. Die Erforschung der tertiären Pflanzenwelt, ihre Methoden, Ergebnisse und Probleme (Senckenbergiana, Vol. 3, pp. 87–98).

Kräusel, R., 1925: Beiträge z. Kenntnis der fossilen Flora Südamerikas (Arkiv f. Bot., Vol. 19).

Kräusel, R., 1929: Fossile Pflanzen aus dem Tertiär von Südsumatra (Beitr. z. Geol. u. Palaeont. v. Sumatra, No. 11).

Kryshtofovich, A. N., 1932: Past Climates of the North Polar Region (In Russian; Priroda (Nature), No. 4).

Kryshtofovich, A. N., 1933: Fossil Flora from the Lozwa River, North Urals (In Russian, Eng. summary; Trans. United Geol. and Prospect. Service of U.S.S.R., Fasc. 291).

Kubart, B., 1926: Bemerkungen zu A. Wegener's Verschiebungstheorie (Arb. Phytopaläont. Lab. Univ. Graz, Vol. 2).

KUBART, B., 1929: Das Problem der tertiären Nordpolarfloren (Ber. d. Deutsch. Bot. Ges., Vol. 46).

LÄMMERMAYR, L., 1923: Die Entwicklung der Buchenassoziation seit dem Tertiär (Repert. Spec. Nov., Beiheft, Vol. 24).

LAM, H. J., 1932: The *Burseraceae* of the Malay Archipelago and Peninsula (Bull. Jard. Bot. Buitenz., Ser. III, Vol. 12, No. 3–4, pp. 281–561).

LAM, H. J., 1934: Materials towards a study of the Flora of the Island of New Guinea (Blumea, Vol. 1, pp. 115–159. Reply by DIELS *in* Engl. Bot. Jahrb., 1938, Vol. 68, No. 2/3).

POHLE, R., 1925: Drabae asiaticae (Repert. Spec. Nov., Beihefte, Vol. 32).

SAHNI, B., 1936: Wegener's theory of continental drift in the light of paleobotanical evidence (Journ. Indian Bot., Vol. 15, No. 5).

SAHNI, B., 1936: The Gondwana affinities of the Angara flora in the light of geological evidence (Nature, Vol. 138).

SCHELLENBERG, G., 1925: Die phylogenetische Entwicklung und die Wanderungen der Connaraceen (Engler's Bot. Jahrb., Vol. 60, Nos. 1, 2).

SCHRÖTER, C., 1934: Genetische Pflanzengeographie (Handw. der Naturwiss., Vol. IV, 2nd ed.).

SCHUSTER, I., 1931: Über das Verhältnis der systematischen Gliederung der geographischen Verbreitung und der paläontologischen Entwicklung der Cycadaceen (Engler's Bot. Jahrb., Vol. 64, No. 1/2).

SCHWARZ, O., 1928: Analytische Studie über die Beziehungen der Phanerogamenflora von Arnhemsland (Nordaustralien) (Fedde's Rep. Spec. Nov. Beihefte, Vol. 51).

SEWARD, A. C., 1928: Greenland as it is and as it was (Nature, Suppl. to No. 3099).

SEWARD, A. C., 1931: Plant Life through the Ages (Cambridge; 2nd ed., 1933).

SEWARD, A. C., 1934: A phytogeographical problem: fossil plants from the Kerguelen Archipelago (Annals of Botany, Vol. 48, pp. 715–741).

SHAPARENKO, K. K., 1935: *Ginkgo adiantoides* (Unger) Heer, contemporary and fossil forms (Philippine Jour. Sci., Vol. 57, No. 1).

SHAPARENKO, K. K., 1936: The nearest ancestors of *Gingko biloba* (In Russian; Bull. Bot. Inst., Acad. Sci. U.S.S.R., Ser. I, No. 2, pp. 5–32).

SHAPARENKO, K. K., 1937: *Liriodendron*, the tulip tree (In Russian; Acta Inst. Acad. Scient. U.S.S.R., Ser. I, No. 4).

SIMPSON, G. G., 1929: Past climates (Nature, Vol. 124, No. 3139).

SIMROTH, H., 1914: Die Pendulationstheorie, 2nd ed. (Leipzig).

SKOTTSBERG, C., 1925: Juan Fernandez and Hawaii (Bernice P. Bishop Museum, Honolulu, Bull. No. 16).

SKOTTSBERG, C., 1928: Remarks on the relative independency of Pacific floras (Proc. Third Pan-Pacific Sci. Cong., Tokyo 1926, pp. 914–920).

SOERGEL, W., 1917: Das Problem der Permanenz der Ozeane und Kontinente (Stuttgart).

SOULLER, P., 1925: Le relief de la terre (Paris).

STEFFEN, H., 1935: Beiträge zur Begriffsbildung und Umgrenzung einiger Florenelemente Europas (Beih. Bot. Centralbl., Vol. 53, Abt. B).

STUDT, W., 1926: Die heutige und frühere Verbreitung der Koniferen und die Geschichte ihrer Arealgestaltung (Dissertation; Hamburg).

SVEDELIUS, N., 1925: On the discontinuous geographical distribution of some tropical and subtropical marine algae (Arch. f. Bot., Vol. 19).

THISELTON-DYER, W., 1909: Geographical distribution of plants (*In* SEWARD's "DARWIN and Modern Science"; Cambridge).

VAN DER GRACHT, W. A. T. M. *et al.*, 1928: Theory of Continental Drift. A Symposium on the Origin and Movement of Land Masses both Inter-Continental and Intra-Continental, as proposed by A. WEGENER (New York).

WALLACE, A., 1880: Island Life (London; p. 81—The permanence of great oceanic basins).

WALTHER, J., 1930: Die Vererbung des Stoffwechsels im Verlaufe der geologischen Perioden (Forsch. u. Forschr., Vol. 6, No. 13).

WEGENER, A., 1922: Die Entstehung der Kontinente und Ozeane (Braunschweig; 4th ed., 1929).

WULFF, E. V., 1934: Essay at dividing the world into phytogeographical regions according to the numerical distribution of species (In Russian, Eng. summary; Bull. Appl. Bot., Gen. and Plantbr., Ser. 1, No. 2).

WULFF, E. V., 1937: Plant geography and WEGENER's theory (In Russian; Priroda (Nature), No. 3).

Chapter XI

CONCEPT OF FLORAL ELEMENTS

An investigation of the origin and history of development of a flora should be based on a study of paleobotanical data and an analysis of the areas of the species of which the given flora is composed. As a result of such an analysis, the flora may be subdivided into definite groups of species. In making this subdivision it is necessary, first of all, to eliminate not only the "aliens" and "escapes" but also the "wides", *i.e.*, species having such an extensive range that they embrace in their area of distribution several phytogeographical regions. Such widely distributed species, called by EIG (1931) "polychores" or "pluri-regional species", cannot give any clues to the origin and history of a flora, at least not unless their centers of origin and time of dispersal have been determined. After these three types of species have been eliminated, the remaining species may be classified into a number of groups, which are termed *floral elements.*

The endemic species in the flora under investigation should first be studied. It would be erroneous to consider that endemic species, since their areas are limited to the given flora, do not require analysis. The establishment of relationships between endemic species, of neo- and paleoendemism, and, in the latter case, the determination of the former areas of these species, as well as the determination of geographical series of vicarious species—all this gives very important clues to the origin of a flora. The remaining species should then be divided into groups according to the geographical character of their areas. This will make clear the regions from which these groups of species originated, and from these data the genesis of the given flora may be established. In other words, on the basis of a geographical analysis of its component species we may establish the origin of any given flora.

The concept of floral elements, as we have defined it, appears so simple that any special explanation would seem superfluous. In fact, however, due to the conflicting interpretations that have been given to this concept, the question assumes considerable complexity. It seems advisable, therefore, to summarize briefly the various points of view.

The concept "phytogeographical element" was first introduced into science simultaneously by two investigators: by CHRIST (1867) in his investigation of the origin and distribution of the flora of the Alps and by ARESCHOUG (1867) in his history of the Scandinavian flora (*see also* DEGELIUS, 1935). But even earlier a number of writers, *e.g.*, WILLDENOW (1802), UNGER (1852), HOOKER (1855), and others, in giving analyses of floras, had in mind precisely such groupings of species, although they did not use the term "element".

The grouping of species within any given flora may differ widely according to the principle upon which this grouping is based. A survey of this problem was first made by JEROSCH (1903) in her book devoted to a history of the flora of the Alps. Many investigations of the same problem have since been made—by SCHRÖTER (1913),

Braun-Blanquet (1919, 1923), Reichert (1921), Eig (1931), Wangerin (1932), Steffen (1935), Degelius (1935), and others. Having made a study of these various viewpoints, we consider that the division of a flora into elements should be based not on any one but on all of the following five principles:

1. *Geographical elements*, species grouped on the geographical principle, according to the types of their areas, according to their geographical distribution within the region whose flora is under investigation, and also, where necessary, according to their altitudinal distribution.

It would be erroneous to believe that simply by grouping the species of a given flora on the geographical principle, *i.e.*, by establishing the geographical elements of a flora, one may determine the origin of that flora. This may be true in part only as regards migration floras, but as regards the relic and endemic species even of such floras and also as regards relic floras, elements established on the basis of geographical distribution do not reflect the genesis of a flora. If, for example, we take arctic-alpine elements, we find that part of their area lies in the Arctic and part in the Alps. But which of the component species are of Arctic origin and spread from the Arctic to the Alps and which are of Alpine origin and spread to the north into the Arctic regions we shall not be able to ascertain by grouping the species on a geographical basis.

2. *Genetic elements*, species grouped according to their region of origin, thus reflecting the genesis of the given flora. To determine the region of origin of a species is often a very difficult matter, requiring a monographic study of the genus to which it belongs. Nevertheless, these difficulties are not insuperable, and the classification, if not of the entire flora, at least of its chief species into genetic elements is fully possible.

3. *Migration elements*, species grouped according to the routes by which they migrated to the given floral region, *e.g.*, along the valleys of certain rivers, along the seacoast, over definite mountain passes, or over detour routes to avoid mountain chains, etc. To establish such elements is ordinarily very difficult and of little promise, since a species may penetrate into the domain of a given flora not by a single route but by several routes. Nevertheless, the establishment of even a small number of such migration elements may provide valuable clues to the history of the given flora.

4. *Historical elements*, species grouped according to the time when they became a part of the given flora (*e.g.*, species that arose in central Europe during the Würm glaciation or species that became distributed during the arid, xerothermic period in post-glacial times).

5. *Ecological elements*, species grouped according to habitat preferences. The establishment of such elements if of great significance in determining the history of a flora and of those climatic changes to which it has been subjected.

The fact that the term "element" is used in so many different senses has impelled some investigators to maintain that it should be used in only one of these senses and that for the others different terms should be applied. Some maintain that the term "element" should be only for designating genetic elements, others that it be applied only to geographical elements. The advocates of the latter alternative advance as grounds the fact that the initiator of this term, CHRIST, applied it only in a geographical sense, and that only later, beginning with ENGLER, was it used in a genetic sense. The advocates of the first alternative maintain that CHRIST used this term in both a geographical and genetic sense. Nevertheless, most investigators have been inclined to believe that the geographical factor is of primary importance and that the term "element" should be applied to it.

This discussion led to a number of proposals of different names for the various categories of elements enumerated above. Thus, REICHERT (1921) proposed to designate as floral elements only genetic elements, and to call geographical elements "components". For migration elements he gave the term "migrants", for those migrants that penetrated into a given flora during the same historical period the term "historical migrants", and for those that arrived from the same general region and by the same route "locative migrants" ("lokativer Migrant").

WANGERIN (1932), in a paper devoted especially to this topic, maintains that the term "element", if used at all, should be used in its geographical sense, but, in view of the confusion that has arisen in connection with the various ways in which the term is applied, he proposes that it be discarded altogether and replaced by an entirely different term, "Arealtype", that would leave no doubt as to its geographical application. In this connection it may be pointed out that this last term is by no means new, for HOOKER (1855) in his "Flora Indica" grouped species into types (European, Siberian, Malayan, etc.), having in view the same concept as WANGERIN.

It seems to us that burdening science with superfluous terminology is not at all desirable; in most cases it does not eliminate confusion but rather contributes to it. The term "component", for instance, is already used in biocoenology and in an entirely different sense. It is much simpler to preserve the term "element" and to add a descriptive adjective (*e.g.*, genetic, geographical, migration, historical, ecological) in order to make clear the precise meaning in which the term is used. This seems all the more advisable, since the new terms proposed do not by far cover all the meanings of the term "element". Thus, the following terms have become widely adopted: relic element, endemic element, Tertiary element, Quaternary element, etc. In order in such cases to avoid the use of the term "element", one would have to devise a whole string of new terms which would never be widely accepted.

Of considerably greater significance are the proposals made by EIG (1931), not so much as regards the concept "element", by which he understands "ecological element", but as regards a more profound analysis of floras, chiefly with respect to ecology and phytocoenology. He points out that, in case the flora of a given territory does not spread beyond the boundaries of a definite floral region, this flora will possess only one *basic element* ("élément base")—he has in mind a relic, not a migration, flora. If, however, the territory under investigation embraces two or more floral regions, its flora will comprise an equal number of basic elements.

Within the limits of a natural floral region (or country comprised within such a region) there may be found territories in which specific ecological conditions predominate, *e.g.*, mountain systems, river valleys, etc., thanks to which in these territories there have found refuge whole colonies of alien plants, called by EIG *inclusions* ("encluses"). The latter may either form independent associations or may join associations formed by the basic elements. In such cases it is necessary to speak either of alien inclusions in the flora or of inclusions in the composition of the basic elements. These inclusions, for the most part, do not have any direct connection with the regions in which they originated. Historically, they constitute remnants of ancient migrations of *mother elements* ("éléments mères"), which gave rise to them.

While in the foregoing case we have to do with elements alien to the given flora and restricted to definite territories, in other cases a floral region may contain a relatively small number of species belonging to alien floras that have penetrated within its limits but are not confined to any narrowly restricted ecological conditions and consequently not occupying any definite territory. Such species EIG terms *penetrants* ("pénétrations"). The occurrence of such penetrants in the composition of any given flora is due, for the most part, to the existence of favorable ecological conditions (*e.g.*, the penetration of a certain number of Mediterranean elements into western Europe without occupying any definite territory there). In many cases these penetrants are likewise remnants from the migrations of ancient elements during periods favorable for such migrations.

Every floral region possesses in its flora, besides its own basic elements and besides inclusions and penetrants, if such there be, a large number of other species more or less widely distributed not only within the limits of this region but in adjoining regions. These species EIG calls *plant links* ("plantes de liaison", "Verbindungs-Sippen"). These adjoining regions have been in juxtaposition since ancient geological times, and they are closely linked by common climatic conditions and other ecological factors. Hence, these plant links are indicators not only of floristic but also of ecological and phytocoenological bonds. It is natural that neither wides ("polychores" in EIG's terminology) nor weeds can be classed as plant links. In some cases it may be found possible to establish the existence of entire groups of plant links ("grouppements de liaison").

Hence, from EIG's point of view, among the autochthonous species of a natural flora we find not only elements of this flora but also plant links and wide-ranging polychores. While we agree that not every autochthonous species entering into the composition of a given flora can be included among its elements, we believe, as already noted, that wides ("polychores") may be so included, provided their centers of origin and time of dispersal have been determined.

Returning now to the concept "element", we wish to consider the question as to which of the above-enumerated types of elements is of most importance or, in other words, on which of them we should chiefly base an analysis of a flora. Some investigators consider genetic elements of most importance; others, such as WALTER (1927), maintain that geographical elements constitute the only reliable basis for subdividing a flora into its constituent parts.

In our opinion neither viewpoint is correct. Geographical elements undoubtedly are fundamental for an understanding of a flora, and an analysis of a flora should begin with these elements. But if we restrict ourselves and do not go more deeply into the genetic significance of these geographical elements, we cannot arrive at any conclusions as to the history of development of a flora. At the same time, the classification of elements solely on the basis of their significance for the genesis of a flora is entirely impossible without a preliminary grouping of species according to their geographical distribution. Hence, it is clear that in analyzing a flora we should not restrict ourselves to either type of elements but should simultaneously study both.

Moreover, the grouping together of species that entered into the composition of a given flora by a common route and the grouping together of species that were enabled to spread during the same historical period are likewise of prime importance for an understanding of the historical development of floras. It is equally of importance to know which species may be grouped together because of like ecological peculiarities that contributed to their entering into the composition of the given flora.

In other words, an analysis of a flora, made with the aim of establishing its origin and the history of its development, achieves its purpose only in case it is carried out on the basis of the subdivision of the flora into *all possible types of elements*. Only by such an analysis, culminating in a thorough, comparative study of all the accumulated data, is it possible to understand a flora, to understand, so to say, its philosophy.

Let us now briefly survey the difficulties involved in referring a species belonging to a flora under analysis to one or another type of element. Thus, as regards geographical elements, it is not possible in all cases to establish the full extent of the area of a species without preliminary study. In many cases a monographic study of a species is necessary in order to define its area precisely. But even in the case of those species whose areas have been established, it is by no means a simple matter to refer them to one or another geographical group. This is clear from the fact that investigators often disagree as regards the geographical distribution of one and the same species. Thus, out of a long list of such disagreements cited by WANGERIN we may take as an example *Dianthus superbus*. OLTMANNS referred it to Atlantic species, *i.e.*, to plants of a humid climate; GRADMANN included it in a group of continental forest plants; HUMMEL considered it to be a Euro-Siberian species; HEGI—Eurasian; STEFFEN—Sarmatian-Central European. The possibility of such diversity of opinion arouses doubts as to the significance of analyses of floras divided into groups on such a basis.

A correct decision as to the geographical element to which a species should be referred may be made only by locating the mass center of the area of the species, for by this means we can determine which part of this particular area is basic. It is quite apparent, however, that this does not make the task easier but rather complicates it. Nevertheless, it is in this direction that studies of areas should proceed, since only in this way will there be excluded any danger of subjectivity in the evaluation of an area and will there be created the possibility of an objective testing of conclusions. Often an evaluation of an area is made without going into the history of the formation of the area itself, the flora of Europe, for instance, being taken as a starting point, with which flora the species of the given flora are compared. Thus, HOOKER (1855) finds in the flora of the Himalayas and northern India a European element only because the species forming it are likewise found in the flora of western Europe, although, of course, it is perfectly clear that the centers of areas of these species lie precisely in the Himalayas, whence in post-glacial times they spread to Europe. Hence, this element might be designated as Himalayan in the flora of Europe but in no case as a European element in the flora of India.

We shall now take up individually some of the floral elements most frequently encountered in analyses of the composition of floras from the standpoint of historical plant geography.

Geographical Elements: —

1. The *arctic-alpine element* embraces a group of species having areas with one part lying in the Alps and the other in the Arctic. These two parts, now separated by the entire breadth of the subarctic zone and the greater part of the north temperate zone, were during the Ice Age in juxtaposition. At present only in a few places do mountain chains stretch from the Arctic zone far to the south, as, for instance, the Rocky Mountains in North America, the Urals in the U.S.S.R., and the Stanovy Mts. in northeastern Siberia, thus forming a bridge between the arctic and alpine parts of the areas of species of this element. By ascertaining the origin of the various species belonging to this element and dividing them into an arctic and an alpine group, we may distinguish between genetic-alpine and genetic-arctic elements.

2. The *subarctic or subarctic-mountain element* (STEFFEN, 1935) embraces species distributed along the border between the forest and the tundra zone. Some species distributed in the subarctic zone are also found in mountains, these forming a subarctic-mountain element. Here we may also refer subarctic steppe species that penetrated into subarctic regions presumably in arid, post-glacial times.

3. The *pontic element* was first introduced as a term by KERNER (1887), who gave it a purely geographical characterization. He understood by it "the very unique flora that spread from the northern and western shores of the Pontic [Black] Sea westward over southern Russia and the northern part of the Balkan Peninsula, the Danube region, Siebenbürgen, and Hungary as far as the Alps and the Carpathians". STEFFEN (1935) has made more precise the boundaries of the range of this group of species. On the south this range begins with the northern spurs of the Caucasus Mts. and the northern shore of the Black Sea, excluding Crimea (which he refers entirely to the Mediterranean region); on the north its boundary lies where the steppe merges into mixed oak forests; on the west this range embraces the Rumanian steppes and the Hungarian lowlands. As regards these three boundaries all investigators are in agreement, but this is not the case as regards the eastern boundary. In view of the fact that species of the pontic group are primarily xerophytes distributed far to the east within the limits of the steppe zone, some investigators have set its

eastern boundary at Lake Baikal; others, on the other hand, consider it to be the Urals. In our opinion, neither boundary is in accord with the concept of "pontic element" in KERNER's sense. Representatives of this element found far to the east do not constitute there a basic element of the flora but merely inclusions or penetrants (in EIG's sense) within the region of entirely different basic elements.

But not only did there arise disagreements as to the boundaries of distribution of the pontic element but the term itself was misused by giving it not a geographical meaning, as KERNER had in mind, but an ecological. "Pontic elements" began to be used for steppe plants in general, xerophytes, etc. In view of the confusion that developed with respect to this term, BRAUN-BLANQUET (1923) proposed to discard it altogether and to substitute the term Sarmatian element. However, both terms were used in various senses, so that the confusion became even greater. For this reason it seems advisable to retain the term "pontic element" and abandon "Sarmatian element".

Considerable confusion was likewise introduced by DRUDE (1890) and later by HAYEK (1923), who divided the pontic flora into two branches, the west-pontic forest flora and the east-pontic steppe flora. To the former they referred the forest flora of Central Europe, which constitutes a remnant of a Tertiary flora that survived the Ice Age in sheltered retreats and then returned to its former habitats; to the latter, primarily steppe plants distributed from the shores of the Black Sea westward over the lowlands of western Europe, which during the Tertiary period lay under the sea. The pontic element actually comprises forest species, since within the limits of the main area of this element there are patches of forest, the species of which cannot be excluded from this element; nevertheless, to refer to the pontic type a purely forest flora is a perversion of the original meaning of this term that only adds confusion to its use.

In our opinion STEFFEN, in the work already cited (1935), has given the most correct analysis of this element. He subdivides the pontic region into two provinces, the *eu-pontic* and the *panonic* (the latter term was first introduced by KERNER, who referred to the panonic element the steppe species of Rumania and Hungary), characterized by corresponding floral elements. Having made a thorough analysis of the species growing in the territory of the pontic region, STEFFEN found a number of groups of species that could not be referred to the pontic element. These include: first, wides distributed in several floral regions; second, species distributed as far east as eastern Siberia; third, pontic-Mediterranean species; fourth, species distributed from the Mediterranean Basin as far as Siberia; and, lastly, plant links—Aral-Caspian and pontic-Caucasian (Transcaucasian).

To this pontic element there has been joined as subelements a number of other elements, which should rather be regarded as independent elements or as plant links. Among such we may mention the Illyrian, to which are referred species of the northern part of the Balkan Peninsula, and the Aquilonarian. The latter term was first introduced by KERNER (1888), who used it to embrace species distributed not only within the limits of the pontic flora but also in the Mediterranean Basin: Spain, islands of the Mediterranean, occasional stations in northern Africa and on the Madeira and Canary Islands. These are floral elements that cannot yet be divided into pure Mediterranean and pontic elements. Out of these elements there arose the pontic elements only after the territory of the pontic floral region was uplifted above the level of the sea and became inhabited by plants, which could have occurred not earlier than the Pliocene (LUDWIG, 1923). Hence, we have here precisely that type of plant links that EIG had in view.

4. Atlantic element. By "Atlantic element" in the geographical sense one understands that group of species whose distribution is, as a whole, restricted to the Atlantic coast region of Europe. Islands lying in the Atlantic Ocean west of Europe and northern Africa are not included in the range of this element. Hence, the main area of distribution of this element is the Atlantic coast of Europe—northern Portugal and Spain, western and northern France, the British Isles, and as far north as the Faroe Islands, western Jutland, and the southern coast of Norway. The humid, maritime climate, with its uniform distribution of precipitation throughout the year and its warm winters, has contributed to the homogeneity of the flora of this part of western Europe.

The Atlantic element may be much more precisely determined than the pontic element, since its area of distribution is not transitional between two other elements. It comprises only a few plant links having connection with neighboring regions. Moreover, the distribution of species of the Atlantic element are, in general, much restricted, due to the predominance at the present time throughout the most of Europe of continental conditions not suitable for these species.

The Atlantic element has been subdivided in various ways by different investigators. STEFFEN (1935, pp. 373–4) has summarized these in the following tabular form:—

I. *Atlantic floral element (in sensu strictu).* The species belonging to this element are either restricted in their distribution to the region of the Atlantic coast of Europe or, if they have spread beyond this region, then only in projecting spurs.

  A. *Eu-Atlantic Group:* Compact distribution in the given region; spurs into neighboring floras very few or none at all.
  (Here belongs TROLL's (1925) *hyperoceanic subgroup*, embracing species growing in the most humid places of the region of the Atlantic flora).

  B. *Sub-Atlantic Group:* Similar to A, but with considerably more extensive spurs into neighboring floras.

    1. *Atlantic-Submediterranean Subgroup,* with spurs extending, for the most part, into the region of the Mediterranean flora.
    2. *Atlantic-Subarctic Subgroup,* distributed chiefly or exclusively in the northern part of the Atlantic coast region of Europe, with spurs extending into the subarctic.
    3. *Submaritime Subgroup,* with spurs in several directions, including regions with a more or less continental climate.

II. *Atlantic-Mediterranean Group,* a group intermediate between the Atlantic and Mediterranean groups, requiring not only maritime climatic conditions but also high temperature.

III. *Atlantic-American Group,* embracing those species of the European Atlantic seaboard found also along the Atlantic coast of North America (identical or vicarious species). These species are, undoubtedly, remnants of a once-united flora, now separated by the Atlantic Ocean, and constitute perhaps the most ancient representatives of the Atlantic element.

Mention should here be made of the so-called "pseudo-Atlantic" species (BRAUN-BLANQUET, 1923), characterized by the fact that, although they are ecologically close to the Atlantic species and are chiefly restricted in their habitats to the sandy strips along the shore and to humid climatic conditions, they have a very extensive range and are, in reality, wides (polychores). This group, therefore, does not constitute a floral element in a geographical sense.

There has been no lack of different interpretations of the Atlantic element. The chief confusion lies in the fact that this term is frequently used as a synonym for "oceanic element" (ozeanischer Element), as comprising species adapted to humid habitat conditions. The oceanic element is a type of ecological element; hence, to confuse it with the Atlantic element (a geographical type) or to make these two terms synonymous, as some investigators do, is entirely incorrect.

The geographical types of elements in Russian phytogeographical literature have ordinarily been designated not by special terms, such as those we have just enumerated, but simply by indicating the geographical regions that served as the centers of origin of these species, *e.g.*, the Altai, Mediterranean, etc. elements. In this way the confusion found in west-European literature was to a considerable extent avoided. Of the few terms applied with respect to the flora of the U.S.S.R. we may mention the following: —

Hirkanic element — species distributed chiefly in the forest and humid regions of northern Iran.

Colchic element — species similar to the foregoing but having their chief center in western Transcaucasia.

Turanic element — species of the Aral-Caspian region, of a desert, xerophytic type.

To devise more special terms is, in our opinion, not of particular service, since such terms are not an aid in understanding the species composition of floras and merely create confusion due to their being variously used and interpreted.

Ecological Elements: — Of the ecological elements those of most vital significance are: the *oceanic element*, embracing groups of species adapted to conditions of a humid, maritime climate; and the *continental element*, embracing groups of species adapted to an arid, continental climate. Of these two chief ecological elements only the first has been well investigated, so that we shall discuss it in greater detail.

Oceanic Element.—In conformity with the very definition of this element as embracing species of a humid climate, it is to be expected that such oceanic species will form part of different geographical elements. All species belonging to the Atlantic element will, at the same time, in an ecological sense, be oceanic elements. It would be entirely incorrect, however, to identify completely the Atlantic and oceanic elements and to use the term Atlantic element in an ecological sense, understanding by it floral elements adapted to a humid climate regardless of their geographical location. Likewise, only confusion is introduced by designating such hydrophytic elements as Iberian, Lusitanian, Macaronesian, insular, etc., when one has in mind not the geographical but the ecological character of these elements. Hydrophytic elements of floras, regardless of their geographical distribution, should be designated as oceanic elements. In other words, there should be included among oceanic elements not only Atlantic elements (in a geographical sense) but also the pseudo-Atlantic species about which we spoke above.

Studies of the oceanic element have been made by many investigators, beginning with WILLDENOW, WAHLENBERG, GRISEBACH, and BLYTT. In more recent years two special works have been devoted to this problem—one by TROLL (1925) on oceanic features in the flora of central Europe and another by DEGELIUS (1935) on oceanic elements among the lichens of Scandinavia.

TROLL distinguishes six different types of geographical distribution of oceanic elements in Europe, which, as interpreted by DEGELIUS, are as follows: —

1. *Eu-oceanic Atlantic type* (*e.g.*, *Vicia orobus*); here belong also such "hyper-oceanic" types as *Asplenium marinum*, *Eriocaulon septangulare*, *Spiranthes romanzoffiana*, etc.
2. *Suboceanic Atlantic type* (*Digitalis purpurea*, *Illecebrum verticillatum*, etc.).
3. *Eu-oceanic Atlantic-Mediterranean type* (*Ilex aquifolium*, *Primula vulgaris*).
4. *Suboceanic Atlantic-Mountain-Mediterranean type* (region of the beech).
5. *Eu-oceanic Atlantic-Subarctic type* (*Myrica Gale*); or *North-Atlantic Subarctic type* (*Cornus suecica*).
6. *Suboceanic Atlantic-Subarctic type.*

This classification, of course, does not by far embrace all the types of species of the oceanic element in the composition of the various floras even of Europe, but it suffices to make clear the distinction between geographical and ecological elements.

The concept "oceanic element" has been subjected to considerable elaboration, thanks to a series of papers by GAMS (1930, 1931–32, 1933), which have made clear not only the ecological character of this element but also its significance in determining the origin of floras. According to GAMS, the oceanic element is represented chiefly by mosses, ferns, and evergreen gymnosperms and angiosperms, *i.e.*, by ancient floral elements. The initial land flora was evolved from water plants that, in the remote ages in the history of the globe, emerged from the water to the surface of the land. The further evolution of this land flora followed, in the main, an uninterrupted course of emancipation from dependence on water and adaptation to continental habitat conditions. This circumstance in itself constitutes grounds for considering oceanic floras as remnants of the most ancient land flora of the globe. It should be noted, however, that not all continental elements are young constituents of the floras of the globe. Climatic zones have always existed, and there were zones with an arid climate and xerophytic floras also in former geological periods. But what is true and what it is necessary to emphasize is that in our time the climate of the globe has become less humid, and the number of regions where flora of an oceanic type may still grow has become ever less and less. As a result, continental elements have become more and more widespread.

Oceanic elements, just as continental elements, may be found in the composition of any flora of the globe. The farther one proceeds inland away from the ocean shores, the more will the ratio between these elements stand in favor of the continental. By selecting from any given flora the oceanic elements and subjecting them to study, we can determine the ratio of these elements to the rest of the flora, the approximate rate of decrease or increase in their number within the limits of the given flora, etc. In other words, by establishing which are the oceanic elements, representing the most ancient elements of the earth's flora, we can obtain data of great value in determining the history and origin of any given flora.

Historical Elements: — Without aiming to give an enumeration of all the groups of species whose distribution is linked with definite historical moments, we shall consider here, by way of example, only two of such elements: *arctic-Tertiary* and *boreal-Tertiary*, terms that were introduced into phytogeography by ENGLER and that have become widely accepted.

According to ENGLER, in our present-day arctic region in the Tertiary period (he, as HEER, considered it to be during the Miocene stage, but it is now referred to a considerably earlier period) there flourished a flora composed of coniferous and deciduous trees now characteristic of North America and eastern Asia (exclusive of its tropical regions). This group of species he called "arctic-Tertiary". Next to this flora on the south there was found the "boreal-Tertiary" flora, bearing to a considerable degree a tropical character, since it comprised palms and evergreen elements. Still farther south, stretching over the vast extent of territory from southern England to Japan and including, on the south, our present-day Mediterranean region, was the domain of a truly tropical flora, which ENGLER characterized as consisting of "megatherms".

ENGLER himself did not give any precise indications as to the changes that took place in these floras during the latter part of the Tertiary and during the Quaternary period. He only indicated, by lines on his map, their migration from arctic latitudes to the Mediterranean region. Many subsequent authors interpreted this in the following way, which undoubtedly is in accord with ENGLER's viewpoint: that all three types of flora—the tropical flora of the Mediterranean Region, the boreal-Tertiary, and the arctic-Tertiary—were forced by climatic changes to migrate southward, each flora replacing that lying next to it on the south.

Paleobotanic data undoubtedly indicate that during the Tertiary period there occurred a succession of floras, tropical and subtropical floras being gradually replaced by temperate and arctic floras, so that the existence of these two historical elements may be considered as practically established. We wish, however, to draw attention to certain drawbacks to the terms applied to these elements. The term "arctic-Tertiary element" does not well express what is intended, since almost to the end of the Tertiary period there was no Arctic zone in Europe. The application of a term employed in modern climatic zonation to the Tertiary allocation of floras makes this designation of the element insufficiently precise, all the more since ENGLER did not restrict the distribution of this flora to the exact boundaries of our present-day arctic region. Similarly, the term "boreal-Tertiary element" does not well express the Tertiary character of the flora which occupied at that time our present-day temperate zone of Europe. In this term the emphasis is placed not on the subtropical nature of those species that we refer to this element but on the present climatic (northern) character of the region inhabited by this flora that subsequently migrated to the south.

But of an even more fundamental character is the following consideration. ENGLER and subsequent investigators presumed that the three floras mentioned succeeded one another completely; they entirely ignored the existence of different floras at different altitudes, with also a gradual dying out of tropical and subtropical elements and their replacement by temperate and alpine elements, in all zones of Tertiary Europe, for which there are ample paleobotanic proofs. Nevertheless, there is no doubt that successions of floras have taken place not only as a result of shiftings of the various elements from south to north but also as a result of a shifting of vegetation belts toward the lower altitudes and the distribution of these temperate and alpine elements in the plains and their migration from east to west or from west to east within the limits of the same climatic zone.

Lastly, it should be noted that the shifting of the tropical and subtropical elements of Eurasia was initiated not from the north but from the east, from eastern Asia, a view to which ENGLER himself inclined in his latest works. From such a viewpoint the term "arctic-Tertiary element" loses its original meaning. In view of their lack of clarity, it would seem advisable to discard both terms—arctic-Tertiary and boreal-Tertiary.

Although it has not been possible to review here all the extensive literature on floral elements, we believe that the principles we have outlined in this and preceding chapters suffice to indicate the general course to be followed in making analyses of floras. This method of analyzing floras is widely used by plant geographers throughout the world. By summarizing the results of their work one may retrace the main stages in the history of the floras of our globe.

References:

ARESCHOUG, F. W., 1867: Bidrag till den Skandinaviska Vegetations Historia (Lunds Univ. Årsskr.).

BRAUN-BLANQUET, J., 1919: Essai sur les notions "d'éléments" et de territoire phytogéographique (Arch. Sci. Phys. et Nat., Vol. 5).

BRAUN-BLANQUET, J., 1923: L'origine et le développement des flores dans le massif central de France (Paris).

CHRIST, H., 1867: Über die Verbreitung der Pflanzen der alpinen Region der europäischen Alpenkette (Denkschr. Schweiz. Nat. Ges., Vol. 22).

DEGELIUS, G., 1935: Das ozeanische Element der Strauch- und Laubflechten-Flora von Skandinavien (Acta Phytogr. Suec., Vol. 7).

DIELS, L., 1910: Genetische Elemente in der Flora der Alpen (Engl. Bot. Jahrb., Vol. 44).

DRUDE, O., 1890: Handbuch der Pflanzengeographie (Stuttgart).

EIG, A., 1931: Les éléments et les groupes phytogéographiques auxiliaires dans la flore palestinienne, I. Texte; II. Tableaux analitiques (Fedde's Repert. Spec. Nov., Beihefte, Vol. 63).

ENGLER, A., 1879, 1882: Versuch einer Entwicklungsgeschichte der Pflanzenwelt, insbesondere der Florengebiete seit der Tertiärperiode, Vols. I and II (Leipzig).

GAMS, H., 1930: *Schisma Sendtneri, Breutelia arcuata* und das Rhacomitrietum lanuginosi als ozeanische Elemente in den Nordalpen (Revue Bryologique, N. S. Vol. 3, No. 1–2).

GAMS, H., 1931: Das ozeanische Element in der Flora der Alpen (Jahresber. Ver. z. Schutz d. Alpenpfl., Vol. 3).

GAMS, H., 1931: Die hygrische Kontinentalität und die Buchen-Arvengrenze (Verh. Schweiz. Naturf. Ges.).

GAMS, H., 1931, 1932: Die klimatische Begrenzung von Pflanzenarealen und die Verteilung der hygrischen Kontinentalität in den Alpen (Zeitschr. Ges. für Erdkunde, 1931, No. 9/10; 1932, No. 1/2, 5/6).

GAMS, H., 1933: Der tertiäre Grundstock der Alpenflora (Jahresber. Ver. z. Schutz d. Alpenpfl., Vol. 5).

GROSSHEIM, A. A., 1936: An Analysis of the Flora of the Caucasus (In Russian; Baku).

HAYEK, A., 1923: "Pontische" und "pannonische" Flora (Oest. Bot. Zeitschr., Vol. 72, No. 9–10).

HOOKER, J. D., 1855: Flora Indica (Introductory Essay) (London).

IRMSCHER, E., 1922: Pflanzenverbreitung und Entwicklung der Kontinente, I (Mitt. Inst. Allg. Bot., Hamburg, Vol. 5).

JEROSCH, M., 1903: Geschichte und Herkunft der schweizerischen Alpenflora (Leipzig).

KERNER VON MARILAUN, A., 1886: Oesterreich-Ungarns Pflanzenwelt (Vienna).

KERNER VON MARILAUN, A., 1888: Studien über die Flora der Diluvialzeit in den oestlichen Alpen (Sitzungsber. Ak. Wiss. Wien, Math.-naturw. Kl., Vol. 97).

KLEOPOV, J. D., 1938: An attempt at a classification of geographical elements as an aid in analyzing the flora of the Ukraine (In Russian; Jour. Inst. Bot., Acad. Sci., Ukr. S. S. R., No. 17, 25).

LOEW, E., 1879: Über Perioden und Wege ehemaliger Pflanzenwanderungen im norddeutschen Tieflande (Linnaea, Vol. 42).

LUDWIG, O., 1923: Das pontische und aquilonare Element in der Flora Schlesiens (Eng. Bot. Jahrb., Vol. 58, Beibl. No. 130).

REICHERT, I., 1921: Die Pilzflora Aegyptens (Engl. Bot. Jahrb., Vol. 56).

SCHRÖTER, C., 1913: Genetische Pflanzengeographie (*In* Handwörterbuch der Naturwiss., Vol. IV; 2nd ed., 1934).

STEFFEN, H., 1935: Beiträge zur Begriffsbildung und Umgrenzung einiger Florenelemente Europas (Beih. z. Bot. Centralbl., Vol. 53, Abt. B., No. 2/3, pp. 330–404).

STEFFEN, H., 1937: Das Pontische Florenelement in Ostpreussen (Schriften Phys. Oekon. Gesellschaft zu Königsberg, Vol. 69, Nos. 2–4).

TROLL, K., 1925: Ozeanische Züge im Pflanzenkleide Mitteleuropas (Freie Wege vergleichender Erdkunde; Festgabe für DRYGALSKI; München).

UNGER, F., 1852: Versuch einer Geschichte der Pflanzenwelt (Vienna).

WALTER, H., 1927: Einführung in die allgemeine Pflanzengeographie Deutschlands (Jena).

WANGERIN, W., 1932: Florenelemente und Arealtypen (Beiträge zur Arealgeographie der deutschen Flora) (Beih. z. Bot. Centralbl., Vol. 49; DRUDE Festschrift).

WILLDENOW, C., 1802: Grundriss der Kräuterkunde, 3rd ed. (Vienna).

INDEX *of* PLANT *and* ANIMAL NAMES

AUTHOR INDEX